Lindicott Colles

Beverly, Mass. 01915

BORN THAT WAY

BORN THAT WAY

GENES | BEHAVIOR | PERSONALITY

WILLIAM WRIGHT

Alfred A. Knopf   New York   1998

THIS IS A BORZOI BOOK
PUBLISHED BY ALFRED A. KNOPF, INC.

Library of Congress Cataloging-in-Publication Data
Wright, William, [date]
Born that way : genes, behavior, personality / by William Wright.
p.   cm.
Includes bibliographical references and index.
ISBN 0-679-43028-8
1. Human genetics.   2. Nature and nuture.   I. Title.
QH431.W785   1998
304.5—dc21   98-6141
CIP

Manufactured in the United States of America
Published June 16, 1998
Second Printing, November 1998

Contents

Preface

MY FOUR-DECADE JOURNEY to this book began when I was an under-graduate at Yale in the 1950s. Psychology was the day's hot subject. Actually, Freudian analysis was the center of heat and light, but the closest a Yale freshman could come to that glamorous, high-fashion truth about human functioning was Yale's psychology department. Undergraduate aficionados of psychoanalysis had to content them-selves with snap-remedy films like *Spellbound*; *Now, Voyager*; and *The Snake Pit*.

Undaunted, I went to New Haven with the intention of major-ing in psychology. If inspired as hoped, I would proceed to medical school and on into psychiatry. I had just read *The Brothers Karamazov* and was enthralled by the saintly brother, Alyosha, who couldn't do enough for everybody. With hopes of emulating his compassion, I envisioned myself an analyst ministering to the broken and tormented. In the process, I might gain insight into my own psyche—which at the time seemed to be careening in directions I didn't like.

Rather than inspired, however, I was depressed by Yale's entry-level psychology. The courses were arid, dull, and obsessed with rats. It was as though Yale hoped, by boring neophytes silly, to weed out stu-dents keen on kinks and aberrations, or even worse, students keen on self-contemplation. I, of course, was keen on kinks and aberrations, my own and those of others, and began to rethink my academic future.

My disillusion had a more substantial component. I did not have to delve very deeply into the psychological literature to learn the degree to which the field was dominated by behaviorism, the theory that humans are born free of any significant innate programming and can be molded as society desires by systems of rewards and punish-ments. It was made clear to undergraduates that whatever back-porch wisdom we may have brought to New Haven about inherited traits was the kind of bunk we were paying Yale to purge. Behaviorism was where it was at. "Theory" is too weak a word for behaviorism's hold on psychology in the fifties; it was the day's indisputable truth and was embraced by virtually every university in the U.S. and Europe.

The vision of the tabula rasa human did not stop with the psychology departments. All the country's social scientists were in the thrall of Ruth Benedict, Margaret Mead, and other anthropologists who promoted the same vision of unprogrammed humans. For decades their disciples saw toddlers as just so much random-access memory onto which any number of operating systems could be downloaded by culture. Generations of anthropologists were happily certain that humans per se were fine; culture and society had turned us into the messes we were. It followed that *different* cultures and societies (and political systems) could refashion the human into something quite adorable. As with the behaviorists, the views of these cultural determinists prevailed.

I felt that the model of the human as a blank slate on which anything could be written was too simple, too tidy, too optimistic. I was, after all, at the institution that had turned Cole Porter into a bulldog and crafted the unpromising raw material of Dink Stover into a *Yale man*! Yes, in their eyes the environment could do anything. (While reading this book, you must forget about the environment as clean air and toxic rivers; psychologists had the word first and use it to mean all the external influences at work on an organism, from womb conditions to nursing home.)

None of this born-free dogma, sometimes called environmental determinism, jibed with what I had experienced in eighteen years of sly observation of my fellow humans. I had known brothers raised in the same family who were entirely different in personality and temperament. My sister and I, growing up in one household with the same parents, the same diet, the same rewards and punishments, were highly dissimilar people. I had read enough about other cultures to know that, Mead's Samoan utopia notwithstanding, certain nasty human characteristics seemed to turn up in culture after culture with disheartening regularity. The social sciences in the 1950s were addressing a human quite different from the one I knew.

My presumption, however, ended with my skepticism. I decided it would be daft to enter a field the fundamental tenets of which I rejected. I wished the psychologists well with their rat mazes and their behaviorist certainties, and headed into the less insistent world of art and literature. (I should not make light of the "rat-runners," although, to my shame, or perhaps to Yale's, I considered this important and fascinating research ridiculous at the time.)

In the 1960s, a decade after I finished college, books appeared and caused a stir that pointed to a very different view of human nature from the one that had driven me from psychology. Robert Ardrey's *African Genesis* and Konrad Lorenz's *On Aggression* took the strongly Darwinian view that humans, like all animals, were products of evolution and, like all animals, were born with an elaborate structure of behavioral dispositions transmitted through the generations by genes.

Simple and logical, even obvious, as this view might now appear, it was considered blasphemy in some quarters—and still is to an immovable few. Large numbers of thoughtful people, however, saw it as important news; a handful even recognized the idea's epoch-shifting potential. The behaviorist Bastille had, if not fallen, suffered a telling blow. This alternative view excited me mightily, not because I relished Ardrey's notion that we humans were domesticated killer apes, but because the basic assumption of inherited characteristics rang true, the entrenched orthodoxy about man-made human nature rang false.

I began reading other books by these authors and the works of scientists with similar ideas—Niko Tinbergen, Edward O. Wilson, Lionel Tiger, Robin Fox, Jane Goodall—and was delighted to discover so many with this Darwinist perspective on our species. My enthusiasm for the changing view of the human was apparently shared by many other nonscientists who may have shared as well my dissatisfaction with the view we all had been fed at college. Whatever needs the books met, they found a wide audience among general readers and attracted broad media attention.

Predictably, social scientists reacted violently to the trend that saw human personality as just another fruit of the evolutionary tree. The reviews of *African Genesis*, predominately written by academics, were mostly negative, some scathing. Like-minded books, all but a few about animals but with clear implications for humans, were similarly trounced, usually in proportion to the degree to which the author suggested a relevance to human behavior. To casual readers, Ardrey and the others were offering an interesting new slant on the human; social scientists and other experts on the human saw the books' ruinous challenge to their systems.

It says something about behaviorism's weak hold on public consciousness that Ardrey's book, and others with a similar outlook, sold so well in spite of the reviews. But their claim to significance was not the large sales—a distinction shared by books about after-death experiences and UFO encounters. Except for Ardrey, the authors were respected scientists with solid credentials. Konrad Lorenz would soon win the Nobel Prize. For all the outcry from the entrenched social scientists, the new books launched a tectonic shift in the established view of human nature, a slow, steady movement that, three decades later, advances in the same direction.

For the next few years I kept up with this literature. Then, in 1980, my enthusiasm for the genetic view peaked when I read an article in *Smithsonian* about a study just then starting out at the University of Minnesota of identical twins who had been raised separately. The similarities of the twins reared in different environments but who sprang from identical DNA was, for me and many others, an unequivocal demonstration of the power of genes to shape personality.

I saw the article as a landmark. The identical-twin study was dragging the infant science of behavioral genetics from behind the safe and remote chimpanzees, geese, and fruit flies and was boldly delivering it at the feet of the animal everyone knew we had been discussing all along: the human. For me, this was an exciting development, too momentous for a magazine article, and I began thinking about writing a book.

As the Minnesota research grew with the discovery of additional separated twins, the scientific community began to take note of the gusher of gene-personality evidence. The popular press took note as well. Although the study was clearly serious and scientifically respectable, it was seasoned to the media's taste with spicy anecdotes about bizarre parallels in the separately reared twins.

Some of the congruences were so improbable they sounded like the imaginings of Gilbert and Sullivan on a bad plot day—for example, sisters arriving at their first meeting each wearing seven rings—and the far-fetched weirdness of the similarities brought down an amount of ridicule on the Minnesota researchers. This was primarily from the antigenetics people who dreaded the implications of Minnesota's more substantial findings and hoped to divert attention away from this data and refocus it on the hard-to-swallow anecdotes. The gene-fearing opposition may have been poking fun, but they weren't laughing.

Understandably, the Minnesotans sought to seal off this vulnerable flank by drawing a veil over the twin oddities. While they were certainly present, and in large numbers, they resisted systematic tabulation, so were useless to the study, yet were undermining the hard-won scientific evidence of genetic influence on personality traits. The result was that quite early in the study, when the Minnesotans spoke publicly, it was only about the statistical evidence of trait inheritance; nothing more was said about the identical hobbies, dress styles, and phobias of the twins who had never met.

I was very let down at this moratorium on the striking twin parallels. I appreciated the psychology profession's comfort with statistics and its discomfort with anecdotal phenomena. It was struggling against a soft-science image. But it seemed that the Minnesota study, with its rare and exotic raised-apart twins, had stumbled on natural phenomena teeming with valid, scientific implications.

I began to see their cautionary silence as an addition to my short list of advantages to a nonscientist writing a book on this subject. Instead of rushing past the bizarre twin similarities as threats to my scientific dignity, I could linger over them, ruminate about them, discuss their implications. Among other pluses I awarded myself: I had no published positions to defend, no scholarly reputation to protect, no academic toes to fear trampling. An overview book for the layman was badly needed and I could approach such a book as a journalist, an infiltrator of unfamiliar worlds, and a translator of the geneticists' arcane terminology.

I even saw advantages to my rudimentary knowledge of molecular biology. For the experts, the amazing chemical interactions propelling all of us are more interesting than the often banal human behavior these systems produce. Adding to the fascination of the molecular level are the newness of this research and the challenge of the fresh enigmas that are emerging on a routine basis. The complex interplay of proteins that influence personality is still a mysterious no-man's-land stretching between genes and behavior. My helplessness in the face of the scientific conundrums would force me to concentrate on the overall cause-and-effect tableaux.

These, then, are the reasons I have for the past twenty years been an armchair follower of this inspiring movement. In addition to my excitement over the subject, I saw the project as atonement for not having stuck with my undergraduate hunches about genes in the first place. Because of this, I dedicate the result that follows to the small band of scientists—such as Irving Gottesman, Sandra Scarr, David Lykken, and Thomas Bouchard—who during behaviorism's brook-no-dissent reign shared my skepticism, but who, unlike me, were not driven from psychology by the prevailing wisdom, but who hung in, courageously, to make their wisdom prevail.

Acknowledgments

WHEN I PLUNGED INTO this alien territory, I was continually amazed by the generosity and patience shown by busy scientists toward a journalistic interloper. Although most of these men and women were consumed with important research, all graciously took time to explain their work to one who—at least at the outset—barely spoke their language. My first thanks go to Thomas Bouchard and his dedicated group at the Minnesota Twin Studies—in particular David Lykken, Auke Tellegen, Matthew McGue, and Margaret Keyes.

Among the many other scientists and science writers whose help stands out in my memory are the following: Jonathan Beckwith, Dorothy Berner, Wade Berretini, Nathaniel Comfort, Barbara Ehrenreich, Frank Elliott, Robin Fox, Irving Gottesman, Dean Hamer, Jerome Kagan, Leon Kamin, Kenneth Kendler, Daniel Koshland, Nancy Pedersen, Richard Pillard, Robert Plomin, Arlen Price, Vincent Sarich, Sandra Scarr, Nancy Segal, Alcino Silva, Lee Silver, Stephen Suomi, Lionel Tiger, Ming Tsuang, Tim Tully, Eric Turkheimer, James Watson, Jonathan Weiner, Jan Witkowski, Robert Wright, and Philip Zimbardo. My thanks to them all.

Special gratitude goes to my editor, Victoria Wilson, who weathered with stoicism and optimism the early efforts of a writer grappling with a new genre. And most of all, I want to express profound thanks to my agent, Helen Brann, who had many solid reasons to discourage my ambition in undertaking this book, but did nothing but cheer me on. Few have such faith.

BORN THAT WAY

ONE

THE CHEMISTRY OF SELF

ON A MONDAY MORNING of a typical workweek, a single woman in her early thirties is awakened by k.d. lang coming over her clock radio. She switches to a classical station and is pleased to hear a Haydn symphony; the soothing rationality of classical music is her preference for starting the day. After a hot shower, she applies her usual discreet makeup, then selects a beige suit with gold buttons to wear to the office. She wore it only four days ago, but she feels good in it and knows it's becoming.

She brews a pot of Yuban French Roast and drinks a cup—milk but no sugar—while skimming the morning paper, which annoys her for having nothing about a major film star's arrest on drug charges that had been mentioned on the eleven o'clock news. An article on the obstructed relief efforts in Zaire upsets her and she resolves to donate another fifty dollars to Save the Children. On her way to the street, the elevator stops at a lower floor and an unfamiliar man in his forties enters and greets her cheerfully. She grunts and feels a complicated mix of obligatory civility, stranger-fear, violated space, and anger at male sexual presumption. She yearns for the elevator to reach the lobby.

Once outside her building, she resists splurging on a taxi and waits for a bus. Finding a seat next to a teenaged boy with books on his lap, one open to a page of graphs, she feels a rush of satisfaction at her success since leaving college, where her grades were lackluster. But then a pang of regret hits her about career opportunities missed, missteps taken. The boy indicates he is getting off. She slides into his seat and looks out the window to see an old man carrying a suitcase stumble on the curb and almost fall. An urge to help comes over her.

The empty seat beside the woman is taken by a tall slender man in his forties. From the corner of her eye she sees that he has curly brown hair, a trim mustache, and glasses—three of her favorite attributes in a male. She reads her newspaper, but glances at his pants leg so close to her skirt and wonders if his leg is hairy. She forces down an erotic surge by plunging into an editorial on redistricting. Traffic is moving slowly; she fears she'll be late. A Hispanic man who looks drunk boards the bus and fumbles for the fare as the traffic light turns red. She wants to scream.

All of these reactions, concerns, judgments, and decisions might seem products of conscious thoughts. Or, with more reflection, some might be traced to the woman's experience, upbringing, or social conditioning. They may, in fact, have sprung from none of these, but may have been prompted, wholly or in part, by her genes, those infinitesimal bits of DNA, which thirty years of research tells us influences our personalities, our behavior, and how we respond to the world around us. As the woman goes about her day, she draws on her reasoning power to deal with special situations. Much of the time, however, she is, like most of us, on semiautomatic pilot, reacting to whatever the environment throws at her with ebbs and surges, blips and flashes, of chemical, gene-rooted responses. As with all humans, her behavior is shaped and guided by signals from the biochemical motherboards that genes have created in each of us.

The interaction between genes and environment is, we now know, essential to the developing child—and for psychologists the term "environment" means every influence on an organism that is not genetic. Not only in children, but in adults too, the environment can have powerful effects. But to a greater degree than ever before realized, the genetic influences on behavior, barring an extraordinary childhood (malnutrition, social deprivation, prolonged abuse), express themselves pretty much as configured before birth.

Scholars have traditionally divvied up the human into an array

4

of discrete vantage points—anatomy, psychology, anthropology, sociology, economics, political science, history. We now see these disciplines converging on a component of our physical selves that mounting evidence indicates is the underlying basis of it all: the twenty-three pairs of chromosomes, containing approximately 100,000 genes, that exist in every human cell. Whatever the term—chromosomes, genes, DNA, the double helix, nucleic acids, ribosomes, alleles—all refer to our biochemical blueprints.

Genetic discoveries have been receiving so much press that nonscientists can be forgiven for seeing the DNA furor as a fad, a New Thing steamroller, this year's channeling or biorhythms. The world appears so in the grip of a double-helix dither that less excitable types shrug it off as a hyped-up media ploy to make the news of the day appear different from the news of last week. Unfortunately for people already bored with gene palaver, the ramifications of this scientific earthquake will continue unfolding, and making news, well into the next century. And the possibilities this knowledge opens up are vast.

Many of the most riveting findings have been of a high-priority medical nature, the genetic roots of birth defects and diseases that, throughout time, have plagued humanity. Because of the widespread suffering caused by these genetic mishaps—British geneticist Steve Jones states that one child in thirty is born with a genetic irregularity of one sort or other—the excitement over genetic therapies is understandable. More recently, the sensational news about cloning a sheep from a mature cell has seized the spotlight from even the landmark medical breakthroughs.

Important as these advances are, they have overshadowed a concurrent, and in some ways more momentous, revolution—the burgeoning understanding of genetic links to personality and behavior. A mass of research that has been building over the past two decades has forced most psychologists and other social scientists to acknowledge what they had long denied: Genes influence not just physical characteristics such as hair color and susceptibility to cancer but our personalities, temperaments, behavioral patterns—even personal idiosyncrasies, the quirks and foibles that make each person unique.

Since behavior is the subject of this book, and the term is broad enough to glaze the eyes of nonprofessionals, it might be a good idea to consider what the word means to scientists. For them, behavior is everything the organism does and thinks—from crying for its mother to delivering a Nobel acceptance speech. Ambition is behavior; so are laziness, rebellion, and compassion. Patriotism, sexism, hating your boss, and loving the Lakers—all are forms of behavior. Virtually anything the individual does, any product of the brain, any action, any mood, emotion, mannerism, or tic, is lumped under the umbrella word "behavior."

From the beginning of the brief hundred years that the mechanics of inheritance have been unfolding, science understood that genes were the building plans for our bodies and brains, the human machine that seemed to be able to think and behave in unlimited numbers of ways. Patterns and constraints were imposed on behavior from the external world, especially from the culture and its primary agents, parents. Now we see that this picture is not accurate. Many patterns and constraints are imposed by culture, but many others, the new evidence shows—along with batteries of impulses, leanings, attitudes, susceptibilities, aptitudes—are born with us.

It is hard enough for nonscientists to conceive of a few microscopic specks of nucleic acid containing the instructions for growing an arm, an ear, or a kidney. Now research says we must grasp as well that similar specks can also go far toward determining if we are to be happy or morose, passive or aggressive, bright or dim, liberal or conservative, religious or atheistic. "Phenotype" is the word scientists use for each genetic manifestation. (Genotype refers to an organism's entire complement of genes, the overall blueprint for each species.) A leg is a phenotype; so are arms, ears, and kidneys. Geneticists have come to consider behavior (or behavioral propensities) just another phenotype.

Among those pursuing this research, some focus on specieswide traits, seeking out the evolved behavioral template shared by everyone. These are the evolutionary psychologists and sociobiologists who try to identify the broad traits that have evolved to make up human nature—aggression, competitiveness, sociality, and altruism would be a few. Behavioral geneticists, on the other hand, are more interested in individual differences. They are ferreting out the genetic influences, if any, that make one person fearful, another bold, one optimistic, one pessimistic, one placid, one fretful. They are also seeking the specific genetic configurations that abundant evidence indicates interact with the environment to produce such common behavioral problems as depression, addiction, obesity, and autism.

An unexpected product of this research is the ever-narrower behavioral nooks and crannies that reveal a genetic component. For most of us it is not difficult to accept at least partial genetic orchestration of broad categories of temperament such as shyness, pessimism, and boldness, among others. Narrower traits, however, such as compassion, extravagance, rule-flouting, and risk-taking, can, without too great an effort of the imagination, also be nominated for biochemical underpinnings. But our minds rebel at the news that genes can induce such behavioral minutiae as hand gestures, pet-naming, and nervous giggles. According to recent research, this appears to be the case. Whether aimed at individual differences or specieswide traits, both behavioral genetics and evolutionary biology are in the business of seeking genetic paths to behavior, and both are bringing about a new perspective on the human complex.

Therapeutic promise is not the only reason the news about gene therapy and cloning has overshadowed the news concerning behavior. The genetic insights about physical defects and dysfunctions are filling a void of knowledge or deepening existing understanding. This sort of information is welcomed by everyone. Findings about the gene-behavior dynamic, on the other hand, are overturning existing truths and demolishing assumptions upon which fifty years of psychological theory has been based. Totally different answers are emerging to questions many experts were confident had long been answered. It is this apostate cast to the behavioral findings that has caused turmoil in the academic community and provoked angry debate. It has also contributed to the early media caution in announcing the discoveries.

THE LARGEST BODY of hard data to establish the genetic roots of behavior has come from comparisons of fraternal with identical twins and comparisons of adopted with biological siblings. For thirty years these investigations have been progressing quietly in scores of kinship studies in the United States and abroad and building a mountain of evidence of the gene-behavior relationship. Of all this research, the most persuasive as well as the most dramatic has been an eighteen-year examination at the University of Minnesota of identical twins who were separated shortly after birth and raised in different homes. The study has examined over seventy sets of separated identical twins and more than fifty sets of fraternals. The telling results startled not only the scientific world but the Minnesota researchers themselves. This was not so much for the degree of genetic influence on traits, which has already been established by other studies, but by the highly specific nature of some genetic expressions. Some of these stories are astounding and dramatically extend the possibilities of genetic string-pulling.

Conclusive as these overall twin findings were for many of the extensive gene-personality links, even harder scientific evidence corroborating this data is just now beginning to emerge from molecular biologists who are tracking the DNA itself to locate explicit chromosomal segments that lead to particular behaviors. While twin and adoption studies measure and compare individuals to establish genetic influence, molecular biologists can be seen as approaching human behavior "from the other end," seeking out the individual genes that might contribute to a particular trait. Such gene pinpointing brings us much closer to interventions—enhancing, fixing, blocking. The possibilities are endless.

After a disheartening series of near misses—studies that appeared to have located culprit genes for specific behaviors but that could not be replicated—success appeared to have arrived in January 1996. Researchers in two different research groups (one in the U.S., another in Israel) isolated a DNA fragment that was consistently longer in subjects who showed a taste for risk-taking. This appeared to be the first time a precise strand of nucleic acid was linked to a particular personality trait. The discovery was reported on the front page of the *New York Times*. It looked as if science had at last "seen" the bouquet of molecules that causes certain people to shoot rapids and bungee jump. As so often has happened with efforts to pinpoint behavioral genes, this finding was cast into doubt by a third study in Finland, but few in the field doubted that such specific pairings of genes and behavior were far off.

Confusion has resulted from the two different methods of linking genes to behavior when misfires of one were mistakenly seen to negate the successes of the other. A typical example of the misunderstanding appeared in a 1994 *Time* magazine cover story, "Genetics, the Future Is Now." The piece was an authoritative overview of the medical discoveries and a discussion of the social and ethical issues they evoke. The article stumbled, however, when it touched on behavior. "Studies claiming to have found genes for alcoholism," it read, ". . . have not held up under scrutiny, but many people still assume such complex behaviors may be predetermined by heredity." With the word "assume," *Time* renders a belief in a genetic component to traits such as alcoholism little more than a hunch. And by the use of the word "predetermined" rather than "predisposed," *Time* sets up a straw-man theory of all-powerful genes held by no behavioral geneticist.

Twin and kinship studies have for decades established genetic components to some forms of alcoholism, depression, and many other behavioral irregularities. To accurately sum up the state of behavioral genetics research at that time, they could say that, no, scientists had not yet found the specific genes, but, yes, they knew that genes were involved. But that was knowing plenty, enough certainly to justify the millions of research dollars that have been spent in the years since seeking out specific genes. The failure to pinpoint gene-behavior tie-ups in no way weakened the hard statistical evidence of inherited predispositions from *somewhere* in the genome.

In the two years since the *Time* article appeared, there have been major advances in gene-behavior understanding, and specific genes for specific behaviors have now been isolated. But the confusion lingers on. Reviewing Philip Kitcher's book *The Lives to Come* in *The New Yorker* of February 12, 1996, John Seabrook wrote: "What if it turns out that there really are genes that influence intelligence, along with a variety of behavioral characteristics . . . ?"

To write such a sentence in a respected publication in 1996 is akin to someone writing in the *New York Times*, "What if it turns out the sun really is the center of the solar system . . . ?" Although the conclusion of genetic influence over behavior has been, to a degree, inferential, the abundance of data has long since removed the issue from what-if-land.

It is understandable that the DNA scanners have created greater excitement than the observational studies of twins and adoptees. The main reason for this preference is that zooming in on trouble-causing genes brings science much closer to therapies and remedies. But those of us rooting for broad acceptance of the new knowledge about genes and behavior see a public relations advantage as well. Astronomers might prove mathematically that Pluto is out there circling the sun, but for most people, it is much more convincing to see a photo of the elusive planet, however fuzzy.

When scientists hold a news conference to show on a screen the gene that causes alcoholism—or depression or violence or nail-biting or child abuse—the impact on the public is sure to be far greater than the hardworking psychologist, who may have examined thousands of twins pairs, holding up charts, saying, "See, my statistics prove a genetic influence." However powerful the numbers, this is still an abstract, circumstantial case; juries always prefer concrete, eyewitness cases.

In animals a number of genes have been located that govern specific behaviors. Most recently a gene in female rodents was found that when blocked, turned mothers from busy nurturers into indifferent loll-abouts. The experiment had clearly isolated a good-mother gene. (Don't have kids without it.) In humans, similar tie-ups are imminent. Since late 1994, several genetic mechanisms that cause obesity have been revealed, and many more gene-behavior couplings are sure to follow.

As the spotlight is turning from the kinship studies with their dry statistics to the sexier molecular biologists with their promise of behavioral-gene snapshots, it should be pointed out that the searches would not have been undertaken in the first place if it were not for the psychologists who for the past thirty years, in bold defiance of their field's prevailing orthodoxy, have been following their intuitions and searching out genetic links to behavior using the only tools at their disposal, the statistical comparisons of genetically related individuals.

Just as Gregor Mendel only inferred the existence of genes but nonetheless developed his on-target laws in the 1860s by merely observing patterns of inheritance, so the behavioral geneticists have drawn conclusions from their studies of twins and adoptees with no idea *which* genes brought about the all-too-evident behavioral effects and little idea *how* they functioned. They pushed ahead, making no secret of their ignorance of the workings in "the black box" that sat between a gene and a trait. Although much mystery still obscures the chemical path from genes to behavior, molecular biologists have joined behavioral geneticists to demonstrate beyond any doubt that the genes-behavior paths are there.

ON THE FACE of it, all of this may not seem too revolutionary. Everyday chitchat abounds in behavioral geneticist thinking. "She got her extravagance from her mother." Or, "the musical talent comes from his father's side." Or, "he's a crook just like his granddaddy." Whether consciously or not, such remarks suggest DNA strings that lead to spending sprees, piano-playing, and crookedness. While we may have gut feelings of genetic transmission of personality traits—animal breeders have known about it for centuries—such thinking has been abhorrent to prevailing scientific thought for much of this century. Back-porch philosophers and animal breeders could believe whatever they wished; science *knew* we humans were creatures of our rearing environments. Experience and learning determined who we are, nothing else.

This is no minor artifact of intellectual history to be tossed quickly into the bin marked "Earlier Mistakes." The behaviorist belief in an all-powerful environment has for many decades dominated enlightened thinking and been the basis of our society's approaches to child-rearing, education, social dysfunctions—and, of course, psychological problems. All the leading psychotherapists, from Freud to Joyce Brothers, might have disagreed about *which* environmental influences made you wet your bed, bite your nails, expose yourself—but none had any doubts that it was *something* in the environment. Always the environment. And this view is still very much with us, if more as a habit of thought than a conscious idea.

Psychiatrist Peter Neubauer, who did his own study of reared-apart twins, tells a story of identical brothers that vividly illustrates the degree to which the environmental assumption dominated our thinking. The brothers, who were in their early thirties, had been separated at birth and raised in different countries. Both were neat and clean to a compulsive degree. When asked to explain how they had become this way, each twin traced his idiosyncrasy to his adoptive mother. One explained that his mother had also been obsessively fastidious, constantly cleaning the house, doing laundry, straightening things. He had no doubt that his tidiness came from her. The other brother, not knowing his twin's explanation, replied without hesitation that he knew exactly why he was so neat: His mother had been a complete slob as a housekeeper. He was reacting against her.

The men did not need training in the day's psychology to have absorbed the behaviorist truth that the key to unusual adult behavior lay somewhere in childhood experience, in the rearing environment. There was no other possibility. Armed with that assumption, explanations were easy to come by, even if they contradicted each other.

In the days when such behaviorist certitude prevailed, countless research projects examined how and to what degree the environment, acting alone, determined behavior. Those skewed psychological studies, which remarkably still go on, were based on the premise that one newborn was pretty much like another; the environment stamped the infants with distinctive personalities. For behaviorists, unraveling the puzzle of a particular human was simply a matter of discovering which environmental elements were the most relevant during the formative years. No one found genetic influences because no one was looking for them. Now that many are looking, they've found them in spades.

The new information about genes is not just a matter of fresh dogma replacing old. The discoveries of behavioral genetics have shown the earlier environment-is-everything model to be half true, but a view of human functioning so myopic, so lopsided as to invalidate most of the findings based on it. It also caused considerable harm. A prime example would be the psychodynamic "cures" imposed on sufferers of conditions we now know can have genetic roots—calamities like schizophrenia, autism, obesity, and an array of neurotic symptoms. Harm was also inflicted when parents were blamed for childhood problems that stemmed from genetic irregularities. In addition to the injustice involved, placing blame in the wrong place moved practitioners further from remedies. For nailing down the cause of psychological problems, genetic knowledge now provides an additional suspect: pesky bits of nucleic acid contained in the genes that the parents may have provided but over which they have little or no control.

A PRINCIPAL REASON the extreme behaviorist, antigenes view dominated psychology for so many years was its strong political appeal. The liberal movements that flourished in the first half of the century based much of their theory on the concept of an infinitely malleable human. The world could be made better by making people better. Talk of genetic influences suggested unchangeable humans and was seen as justification for such societal brutalities as racism and slavery. Not only was a belief in gene-based behavior seen as an obstacle to improving the world, but it was also viewed as an excuse for not trying.

There were valid reasons for fearing a backward thrust to the genetic perspective. Charles Darwin had barely enunciated his theory of natural selection before it was brandished by conservatives as proof of the inevitability of social injustice. His monumental insight about evolution with its encapsulation, "survival of the fittest" (which referred only to procreational prowess), was wrenched into the service of reactionary systems such as Social Darwinism (if you are poor it is because you were born to be poor), eugenics (stop the unfit from propagating), and Nazism (eliminate the unfit already here).

Because of these bogus and preemptive applications of inheritance theories, the entire subject of genes and human behavior was stigmatized with ugly ramifications that linger today. In a look at this history further on, I hope to demonstrate that behavioral genetics knowledge, like all knowledge, is neutral and can be used to bolster any political position, left or right. To reject this fundamental information because of previous right-wing applications makes as much sense as rejecting electricity because of daytime television.

Today, however, the genetic evidence is too powerful for whimsical selection on political grounds of one behavioral theory over another. The abundant data now on the table is forcing both liberals and conservatives to grapple with the genes-environment-behavior nexus. The cumulative evidence from evolutionary psychologists, sociobiologists, and behavioral geneticists has established a new reality with which everyone must deal. Because this revised view of the human is gaining acceptance from people of all ideologies, it is highly unlikely that one group will succeed in commandeering it, at least not without a fight from the others.

The shrinking band who still oppose behavioral genetics on political grounds console themselves with fantasies that it is a passing phase. Prominent psychologist Leon Kamin of Northeastern University, who is one of the most bellicose critics, told me in an interview early in my research that genes-behavior theories make sporadic appearances on the intellectual landscape. With weary exasperation he added that these outbreaks forced him and other alert champions of environmental determinism to beat them back like so many brush fires—writing debunking articles and angry letters to editors. After my five years of research, which I summarize in this book, I feel his remark is akin to saying the spherical-earth theory is an idea that crops up with irritating frequency and must be repeatedly dispatched by scientists of sounder bent.

Political fears are not the only reason for resistance to the genes-behavior findings. Because of broadly held misconceptions about genes' power, many people feel that awarding them a degree of control over our actions means we are no longer masters of ourselves. By admitting that genes influence our behavior, we will be admitting that we are not in charge, that we are operating on chemical remote-control.

The obvious power of genes to dictate physical traits like hair color and foot size has led many nonscientists to assume that if such dictatorial entities also affect behavior, their influence must ipso facto be equally strong in this area. The fear seems to be that if we yield to DNA a degree of sovereignty over our conscious thought, if we permit genes through the behavioral door, we will be acquiescing to the same genetic determination governing our body parts. An individual's chronic pessimism, overeating, or fear of flying would become as unchangeable as eye color.

Behavioral genes don't work that way. None of the data turned up by behavioral geneticists shows genes to be tyrannical commands, but rather nudges, sometimes strong, but more often weak. None of the research has found a single genetic influence on behavior that could be called "all-powerful," even though the critics pin this belief on behavioral geneticists and stigmatize them with the frightening label "genetic determinists."

A lot has been learned in the past year or so about the weakness of some genetic influence. As for the more powerful genetic triggers to behavior—particularly to negative behaviors like addiction and violence—we can now foresee interventions to reduce or alter the effects. It is as though in order to soften the jarring news of how pervasive and meddlesome genes can be, Mother Nature is providing us with calming insights into their uneven power and potential adjustability. With good-news-bad-news timing, we are learning of genes' broad influence over our behavior at the same time we are learning that they are not as powerful as we feared.

Genetic impulses are overruled all the time—when a fat person diets, for instance. As British geneticist Richard Dawkins puts it, when humans use contraception, they are defying their genes, which are bent on reproduction. Nuns and monks take rebellion to the limit and renounce sex altogether. Although the gene-based sex drive is as powerful as any genetic command, it can be disobeyed. However difficult or uninteresting, celibacy is still an option.

While genes are not all-powerful with behavior, evidence mounts they are *all-pervasive* in that they appear to influence, to however small a degree, our every thought and action. This sweeping claim is not just a verbal trick based on the physiological fact that genes have fashioned the brain with which we think and act. It is said in the more specific sense that human impulses, reactions, dispositions, desires, aversions—most facets of our personalities—are colored directly, if only to a minor degree, by each individual's genetic makeup. In countless studies aimed at sorting out genetic from environmental effects, not one of the numerous traits examined failed to show at least some degree of genetic influence.

I have my own hunch why the loss-of-sovereignty fear may linger longer than the evidence should allow. The people most likely to reject the notion of a behaviorally programmed human brain are the brightest people among us, the brain-proud, those most disposed to believe their every thought flows from pure reason, nothing more. They are also the people who evaluate intellectual trends—writers, academics, and other opinion shapers. It is ironic that those who probably see themselves as having the most to lose by admitting that their behavior has genetic strings attached are the very ones we look to for a thumbs-up or thumbs-down on new visions of human makeup. *You* might not be upset to learn that genes are manipulating us humans, but Harvard's Stephen Jay Gould hates the idea.

All of these perceived threats—to progressive politics, to scientific orthodoxy, and to self-esteem—have combined to hinder acceptance of the genes-behavioral revolution and to mute, sometimes to a whisper, the announcements of its remarkable findings. Of course, the validity of the new science in no way depends on our enthusiasm for its implications, but its acceptance, to a large degree, does. Even though the resistance is weakening, the result is that much of the public remains unaware of this landmark change in our understanding of the human or at best perceives the genes-behavior perspective as just one of a parade of fads in psychological truth.

Even worse, the old belief that the roots of all human behavior could be found in the rearing environment permeates much of our thinking and hobbles our quest for solutions to a broad range of problems—from individual quirks to societywide scourges like addiction and violence. Some argue that it's only natural we look to the environment for solving problems because environments can be altered, genes cannot. While both points are untrue, the logic is skewed as well. To use behavioral geneticist Robert Plomin's analogy, this is like the man who lost his wallet in an alley but looked for it in the street because the light was better.

The appealing notion that environments are always adjustable is turning out to be just as false as the notion that genes issue orders we can't refuse. An example of an influential environment that is difficult, if not impossible, to adjust would be the womb conditions that can have lasting behavioral impact on the child to be. Also, the air we breathe and the water we drink—and here the two uses of the word "environment" converge—have elements that can affect behavior surreptitiously, as shown by recent findings about the devastating effects of lead pollution on children. Other troublemaking environmental conditions cannot be repaired because we don't know they're out there making trouble.

In spite of much scientific evidence to the contrary, the two fallacies—immutable genes and malleable environments—linger on. It is, I believe, the shadow of these misconceptions that explains why so many informed people accept in principle a genes-behavior dynamic but have yet to incorporate the new view into their cognitive data banks.

And behind all the pragmatic fears of genetic influence lies an emotional resistance. Will an understanding of the biochemical mechanics of behavior make us less interesting? Throughout history our fascination with ourselves has proved endless. Not only is all of world literature and drama evidence of this, we now have scores of satellites circling the globe, beaming hundreds of channels into millions of living rooms. Except for George Page on PBS and a few other animal programs, all of it is taken up with the rich variations of human behavior. Will an unraveling of behavioral biochemistry, a grasp of the black box, make us less intriguing? I suspect a deep-seated dread that reducing such worthy qualities as friendship, loyalty, ambition, and love to molecular interactions will dispel the mystery, diminish the glory. The resistance to knowledge about behavioral genes may not stem so much from what is implied about politics or self-command but from what it might do to our self-absorption. I like to think it will make us even more interesting, but in a different, more enlightened way. We might even net a few new plots in the human saga.

For all the advances in understanding of genes' power over behavior, no geneticist denies that the environment still plays an important role. Researchers of twins are quick to point out that with separated identical twins, in spite of similarities that so vividly demonstrate genetic effects, their many *differences* are eloquent testimony to the environment's power to mold.

The evidence we are about to examine leads to the conclusion that while the environment *can* have a major effect on forming personality and guiding behavior, much of the time it doesn't. But the simple acknowledgment that, with any human behavior, genes *may* be involved is a momentous change in our species' self-view, and a change with major ramifications for our approach to parenting, education, psychotherapy, and a host of other self-directed human enterprises.

ALTHOUGH SOME ACADEMICS continue to fight for the blank-slate view of the human, the biological-genetic perspective has now established itself in universities throughout the country and, more and more, with the public. Young academic disciplines such as sociobiology and evolutionary psychology are totally rooted in this gene-based view. Increasingly, the traditional social sciences are introducing genes into their grab bags of variables on almost any aspect of human activity. Serious writers are doing the same.

Major behavioral genetics research projects involving teams of scientists are under way at the universities of Minnesota, Texas, Virginia, Colorado, and Louisville, and at Penn State and the Medical College of Virginia. Individual scientists are pursuing smaller studies at Harvard, California State University, and the University of Pennsylvania, as well as at Northwestern and Boston universities. Research at many U.S. hospitals on such medical problems as alcoholism and obesity fall under the category of behavioral genetics. Important research is also in progress in Sweden, Denmark, England, Australia, and Canada.

Increasingly there are signs of gene-behavior awareness seeping into the national consciousness. A *New York Times* crossword puzzle—perhaps as good a bellwether as any other of mainstream acceptance of controversial knowledge—not long ago required a four-letter word that meant "personality determinant." The word was "gene." During a television interview, author Russell Baker was asked by Charlie Rose his opinion about the source of a sense of humor. Without hesitation Baker replied, "It's genetic." Only a few years earlier, few educated people would have held such an opinion, or if they had, would have let it fall so casually from the tongue in front of a large audience.

Now, however, the genes-behavior link has been accepted to such a degree, in both academia and with the public, that opponents are shifting their field to a pose of "what else is new?" While I was talking with Professor Kamin, he launched into a slash-and-burn diatribe against the most prominent behavioral genetics studies. Did he then, I asked, believe human behavior was without genetic influence? With no hesitation Kamin replied, "Of course there's a genetic component to behavior." You would never detect this view by reading Kamin's many attacks or those of his allies.

Other members of the opposition, the stop-genes group some term the "radical environmentalists" and I term the "genophobes," slip similar admissions into their antigene screeds. But these offhand capitulations have a false resonance, akin to a chorus of fifteenth-century bishops after learning of Columbus's voyage, saying, "Of course the world is round. We knew that. Let's get on to something more interesting." To that, one could only say, "Not so fast, gentlemen. For one thing, you'll have to change all your maps."

Whether we welcome or resist a genes-behavior link and whether or not the link is seen to be compatible with this or that political vision, the evidence is now overwhelming that our nature is as much a product of evolution as our physiques, that each of us is born with an array of behavioral dispositions—some noble, some ruthless, some specieswide, some individual, some general, some of a birth-mark specificity—and that the more we know about these internal givens, the more effective we will be at dealing with ourselves, with others, and with the psychological and societal problems that, till now, have proved intractable.

The research that brought us to this new view of the human is fascinating and eye-opening enough to make every one of us see ourselves differently. The nature-nurture war is over, but the way it played out says volumes about political visions' ability to block scientific advances. It also says much about how easily hopes for the way we would like things to be can blind the best intentioned of us to the way things are. Finally, a retracing of this exciting and epoch-defining transformation will illustrate that improving the world will happen not by bending our view of the human to fit our solutions but by understanding the givens we have inherited from our evolutionary past and basing solutions upon them.

TWO

BIRTH OF A STUDY

ONE MARCH EVENING IN 1979, Thomas Bouchard, a professor of psychology at the University of Minnesota, received a phone call from a friend telling him of an article in a local newspaper about a set of male twins in Ohio who had been separated at birth, had lived with no contact for their entire lives, and had found each other two weeks earlier. They were now thirty-nine years old, three years younger than Bouchard. The article stressed the many remarkable similarities between the two men raised by different families.

The friend knew Bouchard would be interested, having heard him discuss the powerful experimental possibilities in raised-apart twins. Bouchard thanked the friend and asked her to drop the story into his mailbox the next day. When he arrived at Elliott Hall, the university's psychology department, he found in his mailbox the clipping plus a second copy left without comment by another colleague. As Bouchard stood in the mail alcove reading the article, he was excited by the account, but unaware that the newspaper story would change the course of his life.

Bouchard, tall, heavyset, with an enthusiastic, outgoing nature and an intense gaze undiminished by the glasses he always wears, was recognized as one of the leading psychologists of his day. For all of his scholarly eminence, he still had the unembellished directness as well as the high-energy gusto of his New England farm forebears. While he projected an outdoorsman's heartiness more than the suavity of an intellectual, he quickly revealed a keen mind, a scholar's exactitude, and a rapacious intellect. This last could be seen in his avid consumption of all the literature touching his field and extended as well to a curiosity about everyone he met. His career choices had revealed a taste for bucking prevailing orthodoxies, but this iconoclasm also surfaced in trivial ways, such as flaunting his enthusiasm for junk food in front of his nutritionally correct associates.

His areas of specialization were problem-solving and I.Q., but for a number of years Bouchard had fantasized about the scientific possibilities of studying identical twins who had been raised separately. To have in his laboratory two individuals with identical genetic makeup yet nurtured in different environments—or better, to have a number of such twin sets—would provide a unique opportunity to establish what effects, if any, genes had on personality and behavior.

He knew the idea for twin-based experiments had been around for a long time. A variation on Bouchard's idea was suggested a hundred years earlier by a British scientist Bouchard admired, Sir Francis Galton, a cousin of Charles Darwin, a brilliant dilettante, and founder of the eugenics movement. As Bouchard pursued the idea, he learned that separated-twin studies had been tried a few times, in both Europe and America, but the studies had only been able to turn up a few pairs, the twins varied in degree of separation (different families but with contact, for example), and the studies were unevenly executed. So the results, which all indicated a genetic role in behavior, were easy to dismiss by the prevailing majority of psychologists who were hostile to data suggesting genes had anything to do with behavior. Still, Bouchard was certain the conclusions were accurate, in spite of their flaws, and knew that one day a twin study would be done that was unassailable.

Over the years, a number of psychologists had been aware of the scientific appeal of such a study. Identical twins result when one egg, impregnated by one sperm cell, splits into two fetuses, hence the term monozygotic twins (MZAs for identical twins raised apart, MZTs for those reared together). This would be in contrast to dizygotic, or fraternal, twins (DZs), who result from two eggs that have been impregnated by two sperm cells. Monozygotic twins have not just similar but identical DNA, while dizygotic twins share roughly 50 percent of their genes, as do siblings born at different times.

There had been a number of large twin studies of a different sort: comparisons of monozygotic twins with dizygotic twins, all pairs raised in the same household. The assumption was that if in measuring various personality traits, a greater similarity was found in identical twins than fraternal, the difference could be attributed to their genetic uniformity, since the rearing environment was the same for each set. In all of these studies, the identical twins proved to be more similar than the fraternal.

Bouchard found these studies persuasive but knew others didn't. The results had been attacked by mainstream psychologists who were wedded to the theory that experience and training were the sole molders of personality. These men and women, called "environmentalists" in the profession before that term came to have a totally different meaning, were assiduous in their efforts to discredit any data that granted a role to genes in the formation of personality. Their principal charge was that identical twins were more similar because of social (that is, environmental) pressures on them to *be* more similar (a complaint later refuted by more than one study designed to detect this effect).

Also, the studies were unwieldy, requiring large numbers of twins for statistical reliability. There were other weaknesses to this method, all of which would be eliminated, Bouchard believed, in a study of separately raised twins. He grew more and more convinced that if done correctly and with enough twin sets, the evidence of gene influence would be irrefutable. Increasingly he considered undertaking such a study himself and examined the earlier studies carefully to see why they had failed to convince the scientific community. Behind Bouchard's conviction of the accuracy of the twin studies' conclusions was a personal belief in a genes-behavior link. It was the sort of gut feeling that balks at prevailing theory and that time and again has launched breakthrough research.

He carefully studied the only American separated-twin experiment, a 1937 study by H. H. Newman, F. N. Freeman, and K. J. Holzinger that was based on nineteen sets of twins. He also examined a 1965 study of twelve sets of reared-apart twins conducted in Denmark by Lars Juel-Nielsen. One in England, involving thirty-seven sets, was conducted by the British psychologist James Shields in 1962. With each of these studies, the environmentalists expended considerable energy in debunking their findings.

These efforts culminated in a 1981 book, *Identical Twins Reared Apart,* by Susan Farber, a psychologist at Columbia University, that painstakingly critiqued each of the twin studies. Spotlighting their flaws, Farber concluded that none of the results could be trusted. The most devastating criticism was that of all the sets of twins examined in the different studies, only three pairs could legitimately be considered reared-apart; the other pairs had contact of one sort or another that, according to Farber, disqualified them as reared-apart. Also troublesome to Farber, some twins had not been separated immediately after birth but had spent many months together as infants. She saw this as another possible corruption of the experiment. Citing other flaws and methodological vulnerability, Farber concluded that all of the studies should be dismissed.

Bouchard felt her verdict was unjust. She saw no significance that the studies, for all their imperfections, arrived at the same conclusion: a degree of genetic influence over behavior. Bouchard also felt Farber demanded an impossible-to-obtain standard of "separateness." Above all, he felt that her book, if fair and accurate, proved only that the earlier studies were not airtight. But he knew she intended more than this. The strong implication of the Farber book was that reared-apart twin studies had their chance and they had told us nothing; the environment still reigned supreme.

For Bouchard, her conclusion's illogic only strengthened his dream of a thorough, flaw-free reared-apart twins study. He was certain that if a sufficient number of pairs could be found, if separation occurred very early in life, if care was taken to include only twins who had had no connection after separation, if they could be studied immediately after reunion, and if their personalities could be assessed by widely accepted personality measures, the results would go far to convince the scientific community of genes' relevance to personality.

One earlier study presented a particularly thorny problem to Bouchard's vision. It had been done by the eminent British psychologist Sir Cyril Burt, who claimed to have studied fifty-three sets of separated twins in the 1930s and 1940s. The results, according to Burt, showed a high degree of heritability to intelligence. At first Burt's findings came under the same sort of attack as the other twin studies; but after Burt's death, the attacks switched to Burt himself. Methodological irregularities and statistical oddities were magnified into accusations of falsified data. To expunge Burt's conclusions of genetic influence, the environmentalists were succeeding in expunging the reputation of Britain's most eminent psychologist.

These charges against Burt, which were later found to be unfounded (and will be examined more fully in chapter 14), offer a chilling illustration of the lengths to which the antigenes forces were willing to go to quash evidence of a genes-behavior connection. What struck Bouchard most forcefully in reviewing this history was that Burt's I.Q. findings had been replicated by three other separated-twin studies: one in America, one in England, and one in Denmark. But this replication did little to save Burt once the environmental forces had him targeted. The Burt drama was the largest scandal to ever rock the psychological profession and was well known to Bouchard, as it was to every psychologist. Whether one gave credence or not to the fraud charges, many recognized the ideological motivations behind them.

Bouchard knew well that he was itching to enter perilous scientific terrain. But his apprehensions were subsumed by his conviction that the psychology profession, his chosen field, was ignoring a major component of human personality and behavior. He also believed that a separated-identical-twin study, for all the imperfections of the earlier studies, was still the most potent method for revealing gene influence. With an assiduous effort to avoid the flaws of the earlier twin projects, he was certain such a study was capable of unlocking fundamental secrets about human nature. It would also force those who insisted the environment determined everything to rethink this bedrock assumption.

Although Bouchard had been trained as an environmentalist, he had long had misgivings and watched with satisfaction as the evidence of genetic influence on behavior accumulated in the late sixties and early seventies. In spite of the opponents' speedy refutations of most of this evidence, he knew that others shared his misgivings and that the orthodoxy of strict environmentalism was weakening its hold on many of his colleagues. Bouchard saw twins separated at birth as the ultimate and much needed nature-nurture experiment.

As his thoughts centered increasingly on setting up such a study, he invariably stumbled on one problem: finding a large enough number of raised-apart twins. Were there many? How would he find them? Persuade them to cooperate? Impossible as it seemed, he clung to the idea. Somewhere there were humans walking the earth at that moment who held the answer to a question that had plagued psychologists and thinkers for centuries. They would be worth finding.

Bouchard discussed strategies for locating reared-apart twins with a colleague in the Minnesota psychology department, Professor Auke Tellegen. They both admitted they had never heard of such twins or even read about them in the press. They believed, however, that reared-apart twins must exist and they pondered ways to find them. They considered placing ads in the *New York Times* and the *Los Angeles Times* but dismissed the scheme as too costly and reaching too narrow a population sample. (Even if the ad turned up enough twins for a study, the opponents had a built-in tool for dismissal: all of the twins were *New York Times* readers and therefore an atypical sample.)

Reluctantly, Bouchard put the idea from his mind and pursued other projects. Several years later, two of his friends, on seeing the news story about the separated Ohio twins, remembered his interest.

STANDING IN FRONT of his mailbox reading about the remarkable similarities of the twins, both of whom were named Jim, Bouchard's hopes for a separated-twin study returned in force, although he knew that one set of twins would have no scientific significance regardless of their similarities. Still, he found the pair too intriguing to pass over and resolved to lure them to Minneapolis with the idea of examining them, as clinical case histories, if nothing more. He contacted the woman who had written the newspaper article, explained his mission, and obtained their phone numbers. The Ohio twins, still in the euphoria of their reunion, quickly agreed to come to Minneapolis to be prodded, examined, and questioned for a week. On March 11, 1979, a month after reuniting, the two Jims reported to Bouchard's office in the University of Minnesota's Elliott Hall.

Of the more than seventy sets Bouchard would eventually examine, many had a sizable number of astounding similarities, but none as many as the Jim twins. In fact, several sets of twins had only a short list of noteworthy similarities. While such lackluster sets were few, had

Bouchard stumbled on such a pair at the outset, it might have damp-ened his interest in the potential of reared-apart twins to reveal secrets of personality formation. The Jim twins, on the other hand, with their boggling list of similarities, would probably have ignited a genetic fire in the most dogged environmentalist.

Enthusiasm, however, was never a big problem with Bouchard. An even more important benefit of the Jim twins' amazing list of parallels was that they drew considerable press attention. This exposure, in turn, brought forward other sets of separated twins. So Bouchard's good fortune was finding at the outset a set of twins who illustrated beyond his wildest imaginings that his genetic hunch might be correct. At the same time, they were instrumental in making feasible a broad study, one that would eventually grow to the most exhaustive and com-prehensive examination of reared-apart identical twins ever done.

APPROXIMATELY THIRTY-SEVEN DAYS after their birth, Jim Springer and Jim Lewis had been adopted into blue-collar families and had grown up eighty miles apart in Ohio. As adults, both had worked hard to maintain modest middle-class lives; each of them, in fact, at various periods in their lives, had held down two jobs. At the time of their reunion Jim Lewis was a steelworker, his brother a records keeper for an electrical company.

When Jim Lewis was six, his adoptive mother told him he had an identical twin brother. He was intrigued but made no effort to find his sibling. Springer also had been told he had a twin. Although he had an urge to take some action, he had no idea what action to take. A friend once told Springer of having spotted him in a bowling alley in a nearby town. Knowing he had not been near the town at the time, Springer had an eerie sensation. He got in his car and drove to the town with the vague hope that fate would bring him to his twin. On a similar impulse, Jim Lewis once traveled to the town of Bradford, Ohio, where he knew he had been born, with the confused hope of running into his twin.

When he was thirty-seven, Jim Lewis, after a talk with his mother about the circumstances of his adoption, resolved to find his twin. He went to the Bradford courthouse and explained to the records keeper that he wanted to get in touch with his brother, from whom he had been adopted apart. The clerk found the information, then phoned Jim Springer's real mother for her permission to release it. Getting her consent, she gave Lewis his brother's number. He was stunned that it had been so easy.

When the brothers finally spoke on the phone, both were full of emotions, but, more than anything, they were wary and apprehensive. They chatted for a while and concluded their conversation by making a date to meet at Springer's house on the outskirts of Dayton. "By the way," Springer asked, "What do you drink?"

"Miller Lite," Lewis replied.

"So do I."

As the agreed-on time approached, both Jims grew increasingly nervous. They later admitted to each other their fears: What if the other twin was a bad guy? A jerk? A wife-beater? Needed money? Lewis's anxiety fired his imagination: What if his brother needed an organ transplant? Like sleeping dogs, unknown twins might be better left alone.

Once together, the tension disappeared quickly. While there was a clear resemblance—both were nice-looking, dark-haired men with sensitive eyes and medium builds—the physical resemblance was less striking than in most identical twins, more like the similarity of fraternal twins. The initial impact of their obvious twinness came more from the way they moved, held themselves, spoke, and gestured. As often happens with reunited twins, both felt an immediate bond that went beyond mere rapport and affection but was more like the restoration of a long-missing piece of their lives. In the instantly relaxed and happy atmosphere, they began discussing themselves and the astounding list of similarities began to unfold.

The parallels in the lives of the Jim twins has been told many times—in magazine articles, on television, in lecture halls, in scientific papers. They have been discussed with amazement in laboratories at M.I.T., in faculty lounges at Oxford, and in the *Tonight Show* production offices in Burbank. The catalog of their similarities has become a litany of the behavioral genetics faith. Their dramatic story would bring broad attention to Bouchard and the study he hoped to launch, but their similarities would also come back to haunt him because of the many hard-to-swallow oddities and the resistance of these oddities to scientific explanation. Because of the Jim twins' importance to the

Minnesota study, and to behavioral genetics in general, their parallel quirks and traits bear repeating yet once more.

IN SCHOOL BOTH JIMS had been poor students. One had dropped out in the tenth grade, the other just squeaked through high school. Both had worked as sheriff's deputies, both drove Chevrolets, both chain-smoked Salems, and both like sports, especially stock-car racing. But both *disliked* baseball. They were amazed to discover that they had each taken an impromptu vacation to Florida during a particularly cold winter, had driven their families south, and had selected the same three-mile stretch of Gulf Coast beach as a place to spend their holiday. Coincidence was piling on to coincidence and threatened to collapse into something else.

Some of the oddest similarities involved proper names, as they would with many of the separated twins Bouchard would later examine. Both Jims had married women named Linda, divorced them, and married women named Betty. They were romantic and affectionate, and each had a habit of leaving love notes to his wife around the house. Springer had three daughters and a son named James Allan. Lewis had three sons, one of whom was named James Alan. There was no family precedent in either case, both simply liked the names. Both had owned dogs that they had named Toy.

The two Jims were devotees of the same hobby, woodworking, for which they each had created nearly identical basement workshops, with the same corner placement of the work bench. A highly similar array of tools hung on the walls. Both men passed good portions of their spare time in their shops, each building furniture and picture frames. One of the most striking similarities in the entire Jim-twins catalog was that each had built a circular white bench around a tree in his front yard—although one was made of wood, the other of metal, the benches were unlike anything in their respective neighborhoods.

Perhaps less surprisingly, there were remarkable physiological similarities as well. Both had undergone vasectomies, and both had slightly high blood pressure. Each had become overweight at roughly the same time and had leveled off at the same time. Both suffered severe migraine headaches that lasted approximately half a day and that did not respond to any medication. When they were later interviewed by a psychiatrist at the University of Minnesota, both used exactly the same words to describe the headaches: "Like somebody's hitting you with a two-by-four in the back of the neck."

Understandably, Bouchard was amazed by the number of similarities and their degree of specificity, but he saw no way to establish scientific significance to the phenomena. For all their remarkable parallels, the Jim twins would be seen as nothing more than a bizarre artifact, genetic flukes. At the same time, Bouchard knew the similarities had to mean *something*.

While Bouchard was pondering his frustration, he happened to speak with the journalist who had helped him find the Jim twins. She was curious. What had his interest been? What had he planned to do with them? What had he learned so far? Still excited over his exotic specimens, Bouchard said that while Springer and Lewis were only one set, they were extremely interesting to him as case histories and had rekindled his hope of doing a large-scale study if a sufficient number of separated twins could be found. He discussed the difficulty of finding such twins. The reporter asked how he would finance such a study. Bouchard replied that money would be no problem, he would "beg, borrow, or steal" if he had to.

The reporter worked for a wire service and a few days later a small article based on the Jim twins' similarities and Bouchard's comment appeared in a number of newspapers, including the *New York Times*. Suddenly Bouchard's phone began ringing with requests for interviews with him and his Jims, and for television and radio appearances. Then a call came from *The Tonight Show* asking Bouchard if Springer and Lewis could appear with Johnny Carson. The Jims, who saw all the hoopla as a prolonged celebration of their reunion, happily agreed to everything, and Bouchard phoned their acceptance, but not without apprehension. He knew the jealousies and snobberies that permeated the academic community, and he knew the threat to scientific respectability posed by an appearance on the popular *Tonight Show*, which leaned more toward juggling acts and cleavage than serious science. But he also knew the price of an ad in major newspapers, and he saw a chance to gain a cost-free nationwide appeal for more twins.

Before the Jim twins arrived at the University of Minnesota, Bouchard sat down with two of his graduate students, Margaret Keyes and Susan Resnick, and put together an assessment battery. To avoid being accused later of tailoring tests to hoped-for results, he selected several well-established tests of personality. For the comprehensive series of tests envisioned, he needed the help of other specialists—psychiatrists, medical doctors—so he solicited university friends for assistance. When he approached a professor of psychiatry, Len Heston, to run a psychiatric evaluation on the twins, Heston told Bouchard he felt the undertaking a complete waste of time.

"Len," said Bouchard, "humor me."

Heston went along and later admitted that of all the experts Bouchard recruited into his project, none felt that examining the twins would turn up anything useful. But after a week of examining the two Jims, all were astounded at the degree of sameness and became infected with Bouchard's enthusiasm for studying reared-apart twins.

During the week the Jims spent at the University of Minnesota, Bouchard subjected them to the battery of tests he and his colleagues had devised. In one series designed to measure personality—such traits as conformity, leadership, sociability, tolerance—the two Jims' scores were as close as when one person takes the same test twice. On tests that measured I.Q. and interests, their scores were also very close. Their heart rate and brain-wave measures were almost identical.

Both Jims were cooperative and seemed to relish the attention. This limelight factor would later be pounced upon by critics of the Minnesota study as a possible skewing element. The reasoning would be that twins reveling in the lionization their twinness brought them would emphasize, even exaggerate, their sameness. Astute as the argument is, it is refuted by the separation of the twins during tests and by the types of tests Bouchard administered. They were not only an array of physiological tests but personality questionnaires so complex that there would be no way to falsify or exaggerate. Beyond this, all who were present at the two Jims' first meeting with each other scoffed at the suggestion the two men worked at being similar. Before they had time to collaborate on a "twin routine," and before either had any idea their relationship would interest people beyond their immediate families, the two discovered a list of similarities that astounded them even more than they astounded the others present.

A naturally cordial and accommodating man, Bouchard treated his first set of twins like two extremely rare albino greyhounds that had been placed in his care. (Seventy twin sets later, he would still show the same paternal protectiveness to each pair.) Everything the Jims did fascinated him, and he was constantly vigilant for new astonishments.

Sitting with them at lunch during their week together, he noticed Jim Lewis place his finger in his mouth and bite his nail. Bouchard was intrigued and wondered if by some miracle the brother might have a similar habit. A few minutes later, he watched in amazement as Jim Springer's hand went to his mouth and he began biting his nails. It is unlikely that an undesirable personal habit ever gave anyone as much pleasure as this did Bouchard. Jumping in, he learned that both Jims had been intense nail-biters all their lives.

Bouchard's excitement at watching two adult males stick their fingers in their mouths is understandable. In the behaviorist orthodoxy from which Bouchard came, few things were considered a more classic example of an environmentally induced neurotic symptom. Throughout the thirties, forties, and fifties, millions of nail-biting children were subjected to a bouquet of theories about their habit, ranging from hostility toward parents to castration fears. In the welter of facile and fanciful explanations, no one ever suggested that nail-biting might be an inherited trait. And while the Jims' chewed nails did not prove it was, the phenomenon threw open a long-locked door to the possibility.

Another occurrence struck Bouchard with comparable force. A group of examiners was sitting with Jim Springer playing back a tape-recorded interview and was trying to determine which Jim was speaking. Springer listened for a few moments then said without doubt, "That's me." As it turned out, it was his brother.

While there might be a tendency to lump male speaking voices into general categories like high, medium, and Henry Kissinger, the speed with which an acquaintance's voice is recognized on the phone suggests how distinctive each voice is. That commercial banks and other institutions now trust electronic voice measures as valid identification would suggest that speaking voices are as individual as fingerprints.

An experienced psychologist, Bouchard knew too well how easily most of his colleagues in the field would explain twins raised in the same house having identical speaking voices—the constant togetherness, the subconscious mimicry, and so forth. That these two men had never met until they were thirty-eight yet still spoke exactly the same—the same intonation, the same inflections, similar choice of words—struck Bouchard as highly significant. "Anyone's speaking voice is important behavior," Bouchard would later say. "It's part of your personality."

The twin-recruitment strategy worked. After the *Tonight Show* appearance, a number of twins responded, and within a few weeks Bouchard had five sets and the funds to bring them to Minneapolis for a week. Within another few weeks, five more sets had turned up and expressed a willingness to fly to Minneapolis for a week of tests. The Minnesota Twin Study was born. It would be the first separated-twins study in America in forty-two years. It also went far beyond the earlier studies in terms of extent of testing, the numbers of twins, and the integrity of their separateness. So superior was the Minnesota study to previous studies that various journals, including the *New York Times*, in announcing its launching referred to it as the *first* such study.

Eighteen years later Bouchard's elaborate project was still in progress and had by then examined 120 sets of reared-apart twins, seventy of them identical, fifty fraternal. He and his associates in the twins project, who would total eighteen scientists, had published over forty papers and articles summarizing the results of their investigations in such preeminent professional journals as *Science*, the *Journal of Applied Psychology*, and the *British Journal of Psychiatry*. In the view of many interested and impartial observers, their remarkable findings were the empirical culmination, the clincher, to a mounting pile of evidence of a significant genetic component to human personality and behavior.

In addition, the numerous specific similarities between twins who first met as adults, anecdotal and untabulated as such evidence was (the Minnesota study made no systematic attempt to seek out such particular similarities between twins), strongly suggests manifestations of genetic expression in behavior of a pinpoint exactitude that had never before been so much as conjectured. Despite the oddity and logic-defying nature of many of these parallels—which will be discussed later—the examples are too numerous to ignore and, in many cases, too precise to dismiss as coincidence. If they proved nothing else, the odd similarities would serve to open minds to the possibility of a far broader range of genetic influence over our personalities than had been suspected by even the most bullish geneticists.

But more than the bizarre similarities, it was the array of across-the-board similarities found in a majority of the reared-apart twins that surprised the Minnesota researchers. They had expected to find a degree of concordance in domains like I.Q. and fundamentals of temperament like timidity or gregariousness. They had not been prepared, however, for the identical or highly similar speech patterns, body carriage, sense of humor, temperament, matrimonial histories, professional careers, tastes in clothes, choice of hobbies, and on and on into the labyrinth of personality. Because they were in totally unanticipated areas, many of these similarities turned up accidentally. No questionnaires or tests had been devised to ferret them out and measure them, so hurried efforts were made to include the often repeated categories in the tests. Bouchard had expected a stream of genetically evocative data; instead he was getting a river.

To show how far his original findings were from prevailing psychological thought, Bouchard, in an overview essay on his twin study, quoted a sentence from one of the day's most widely used psychology textbooks, the 1981 *Introduction to Personality* by W. Mischel. After discussing the power of the environment to mold personality, the author writes: "Imagine the enormous differences that would be found in personalities of twins with identical genetic endowments if they were raised in two different families."

THREE

DESPERATELY SEEKING TWINS

AT THE TIME he began the Minnesota Twin Study, Bouchard felt reasonably certain that he would find a high heritability to I.Q., as the earlier studies had. He also foresaw finding a degree of heritability to a number of personality traits, probably the traits that vary least throughout a lifetime, such as timidity, optimism, aggressiveness. While a principal objective was to put to rest the notion that genes play zero role in personality and behavior, Bouchard denies having been on a single-minded quest for genetic influences but rather hoped to bring his profession to a more balanced genes-and-environment view.

"We really intended to look for differences between the twins," he told me in an early conversation. "Whatever differences we found, we knew they had to be environmental. For a psychologist, environmental effects are the most interesting." With the amazing similarities they found in Jim Springer and Jim Lewis, however, they knew that they were onto something quite different, that the genetic effects would be the big news.

Regardless of whether Bouchard felt the objections to the earlier studies had merit—and some he considered downright silly—he made a strenuous effort to address each of them and struggled as well to anticipate unforeseen ones. Twins were tested separately, and well-established tests were used to avoid accusations of customizing the questions, tilting them toward hoped-for results. To forestall charges of examiner bias, outside examiners were brought in to administer I.Q. tests. Although Bouchard told twins the study's purpose, he did not belabor it or go into much detail. (The first set of twins I interviewed after undertaking this book, two young men from a coal-mining town in Pennsylvania, had no idea why they were tested in Minnesota, were mildly interested when I told them, but spoke more of the fun they had had in Minneapolis.) The fear was that twins, knowing what the Minnesota researchers were seeking, would try, consciously or unconsciously, to accommodate them. Most of the twins had little idea whether the Minnesota people were interested in similarities, differences, or neither.

Despite his great caution and the rigorous amount of testing he planned, Bouchard had no illusions of avoiding a hostile reception from the usual quarters if he turned up persuasive evidence of genetic influence—which he felt quite sure would happen if he found enough pairs of separated twins. Bouchard took heart, however, in the intellectual climate's gradual shift then taking place toward an evolutionary perspective on human nature. Four years earlier, it had made a giant move in that direction with the publication of *Sociobiology—A New Synthesis,* by Harvard's Edward O. Wilson. This major work primarily addressed animal behavior but concluded with a controversial bang: a chapter that advanced the notion that in humans too, behavioral traits had evolved along with physical traits. Almost rudely the book said, "By the way, these animals we've been talking about? You're one too."

For this impertinence the book was brutally attacked by the custodians of environmentalism, but it still had a sizable impact with others less threatened and won many converts to the biological perspective. Still, Bouchard knew that there were cadres of environmentalists— now called "radical environmentalists," as the evidence against their extreme position mounted—who would reject his findings no matter how sound his methods. "We knew," Bouchard later said with philosophical resignation, "that no matter what we did, we would still get jumped on."

Bouchard threw himself into setting up his twin study with the added gusto of one who knew he was onto something truly important. He worked long days and nights, reading every scrap of information on the subject and brainstorming with colleagues ever more ingenious strategies for prying cosmic secrets from identical twins. Throughout the planning stage, he was constantly alert for any deviation from methodological impeccability.

The project took on the ambition and scope of a new business venture, and Bouchard brought to it the same drive and resourcefulness of an entrepreneur who has staked all his savings, but in this case his intellectual conviction added a different sort of passion and intensity. When the study reached its eventual scope and magnitude, Auke Tellegen, who now knew the formidable logistical and methodological problems that he and Bouchard had not foreseen when they were fantasizing about a study, paid a simple tribute: "Only Tom Bouchard could have brought this twin study into existence."

BARBARA PARKER, an attractive woman of thirty-six, returned to her Los Angeles home one day in 1983 to be intercepted on her front walk by her next-door neighbor. "Do you remember telling me you were an adopted child and thought you might have an identical twin?" the neighbor asked.

The woman nodded yes.

"Well, brace yourself," the neighbor went on. "She's sitting in my living room."

This was typical of the jarring scenes and intense emotions that typified twin reunions. Some of the twins, to be sure, knew of their twin's existence, as had the two Jims, but had never had the inclination to initiate a search. Others had heard hints from adoptive parents who had learned of the sibling from the adoption agency, but these new parents had either forgotten the information or had decided against telling their adopted child for fear of negative repercussions. Other times an adopted child simply had a strong belief in a lost twin based on no conscious knowledge. Fantasy siblings are common with children, but in the case of Bouchard's reunited twins, the missing-half stories are told with impressive frequency and a high degree of conviction.

A number of the reunion stories hinged on one twin's being mistaken for the other by a stranger, as with the Jim twins. If this sort of chance run-in happened to an adoptee who suspected or knew of a twin's existence, he or she could usually elicit from the stranger enough information to find the other twin. ("This guy who looks like me; do you know where he lives?") This sort of chance encounter has better prospects of success than plunging into court records and other bureaucratic thickets. Documents are often lost and families disperse into untraceability over the years. Even without such mischance, adoption agencies traditionally have had strict policies against releasing information about biological parents.

In recent years, however, this stand seems to be softening, perhaps as a result of a growing awareness of the importance of biological inheritance. In the case of twins, the agencies have heard twin sense-of-loss stories and have come to realize how misguided, even cruel, their earlier policy had been of separating twins—on the hard-nosed assumption that one infant is easier to place than two—and have all but stopped separating twins. As a result, a study like Bouchard's will not be possible in the future.

To TAKE PART in the Minnesota Twin Study, twins had to volunteer, which meant they were a self-selected sample. This is a category in psychological experiment that raises concerns, in that self-selection may not be as pure as, for example, picking a group of people at random from the phone book or, as in other twin studies abroad, locating experiment subjects through national twin registries.

One of the criticisms later leveled at Bouchard's twin study was that because he solicited his twins through talk shows and news articles, he only lured twins into the study who craved publicity and would thus bias his sample. Instead of getting a cross section of American adults, he would get an atypical collection of self-delighted show-offs, who felt themselves on a remarkable level of sameness; dissimilar twins would choose not to compete in what might appear to be a national similarity play-off.

As a group, psychologists have become masterful at ferreting out potential flaws in studies of humans, especially any aspect of the recruitment process that might produce an atypical sample. This is appropriate, since bias in selection methods can creep in inadvertently and must constantly be guarded against. A legendary example of a tainted sample was a major study on criminal intelligence that examined a large number of prison inmates. The costly study went well, with evidence building toward a highly significant result: that criminals were on average of lower I.Q. than the general population. Eventually, some spoilsport pointed out that if the prison criminals measured poorly, it was probably because the smart criminals did not get caught. The results of this particular study did nothing to prove that *criminals* were dumb, only that prisoners were.

As for Bouchard's twins being an unrepresentative sample because they were publicity prone, it is hard to imagine any twins, seeing the Jim twins chatting with Johnny Carson, assuming that they too would be ushered to *The Tonight Show* guest seats if they enlisted in Bouchard's study. It is also hard to imagine twins shying away from the study because they felt they lacked *Tonight Show*–caliber sameness. This analysis ignores the powerful emotions that reunited twins all felt and suggests that their feelings gave way to a desire for publicity.

Bouchard estimates that about 5 percent of the twins he examined had an appetite for publicity. Repeatedly it was demonstrated that the twins' motives in participating was their interest in each other; all other considerations were minor in comparison. In addition, it was never demonstrated that an alleged interest in publicity would affect test results. Of even greater scientific weight than such ad hoc answers to ad hoc criticisms, most of Bouchard's findings were later replicated by a Swedish study that recruited its subjects in a totally different way, one that did not in any way involve publicity.

A matter of equal indifference to Bouchard's twins, it turned out, was the hope of helping tip the nature-nurture debate in the direction of nature. Aside from the twins who were ignorant of the study's objectives, most were oblivious to the genes-environment controversy to begin with and showed little interest in it during their stays in Minneapolis.

For the majority of the twins who volunteered for Bouchard's study, the primary motive for participation was an all-expenses-paid week of togetherness. Once reunited, twins invariably wanted to spend as much time together as possible. Two elderly Scottish ladies were among several sets who ended up living together. Understandably, in the first days after reuniting, twins were obsessed with becoming acquainted, of making up for lost time. In most cases, however, this was not easy. They often lived in different cities, sometimes different countries, they all had the usual entanglements of jobs and families, and meetings could be beyond budgets.

The ages of Bouchard's twins at their first reunion varied widely, but the mean age was thirty. Somewhat greedily, he included twins who were not separated immediately after birth, although the average age at separation was five months. Nor did he exclude twin sets who, although raised separately, had had occasional contact during their formative years. When the critics learned of this relatively rare exception to the twins' reared-apart purity, they pounced and declared it a crippling flaw, a wipe-out contamination of the study's claims of separateness. The reasoning seems to have been that with a pair of twins living apart, one Christmas dinner together could bring such traits as I.Q. and cardiovascular rhythms closer together.

WHEN BOUCHARD'S TWIN-PROCESSING operation was in full swing, he amassed a staff of eighteen—psychologists, psychiatrists, ophthalmologists, cardiologists, pathologists, geneticists, even dentists. Several of his collaborators were highly distinguished: David Lykken was a widely recognized expert on personality, and Auke Tellegen, a Dutch psychologist on the Minnesota faculty, was an expert on personality measuring.

In scheduling his twin-evaluations, Bouchard tried limiting the testing to one pair of twins at a time so that he and his colleagues could devote the entire week—with a grueling fifty hours of tests—to two genetically identical individuals. Because it is not a simple matter to determine zygosity—that is, whether twins are identical or fraternal—this was always the first item of business. It was done primarily by comparing blood samples, fingerprint ridge counts, electrocardiograms, and brain waves. As much background information as possible was collected from oral histories and, when possible, from interviews with relatives and spouses. I.Q. was tested with three different instruments: the Wechsler Adult Intelligence Scale, a Raven, Mill-Hill composite test, and the first principal components of two multiple abilities batteries. The Minnesota team also administered four personality inventories (lengthy questionnaires aimed at characterizing and measuring personality traits) and three tests of occupational interests.

In all the many personality facets so laboriously measured, the Minnesota team was looking for degrees of concordance and degrees of difference between the separated twins. If there was no connection between the mean scores of all twins sets on a series of related tests—I.Q. tests, for instance—the concordance figure would be zero percent. If the scores of every twin matched his or her twin exactly, the concordance figure would be 100 percent. Statistically, any concordance above 30 percent was considered significant, or rather indicated the presence of some degree of genetic influence.

As the week of testing progressed, the twins were wired with electrodes, X-rayed, run on treadmills, hooked up for twenty-four hours with monitoring devices. They were videotaped and a series of questionnaires and interviews elicited their family backgrounds, educations, sexual histories, major life events, and they were assessed for psychiatric problems such as phobias and anxieties.

An effort was made to avoid adding questions to the tests once the program was under way because that meant tampering with someone else's test; it also would necessitate returning to the twins already tested with more questions. But the researchers were tempted. In interviews, a few traits not on the tests appeared similar in enough twin pairs to raise suspicions of a genetic component. One of these was religiosity. The twins might follow different faiths, but if one was religious, his or her twin more often than not was religious as well. Conversely, when one was a nonbeliever, the other generally was too. Because this discovery was considered too intriguing to pass by, an entire additional test was added, an existing instrument that included questions relating to spiritual beliefs.

Bouchard would later insist that while he and his colleagues had fully expected to find traits with a high degree of heritability, they also expected to find traits that had no genetic component. He was certain, he says, that they would find some traits that proved to be purely environmental. They were astonished when they did not. While the degree of heritability varied widely—from the low thirties to the high seventies—*every trait they measured* showed at least some degree of genetic influence. Many showed a lot.

For the most part, the twins accepted the rigorous routine in good spirits, but occasionally one rebelled at the grueling pace. Returning from a hurried lunch to an I.Q. test, an elderly Australian lady complained angrily that she had not been given time to finish her coffee. Then, confronting a particularly tricky spatial-relations puzzler, the woman threw down her pencil and refused to proceed, snapping, "I can't do this bullshit."

When told of the outburst during a staff meeting Bouchard let me attend, he turned to me and whispered, "It's not like working with white mice."

By the end of the first year, Bouchard had examined twelve pairs of reared-apart identical twins and more pairs continued to come forward. Looking back on this period he says, "I got greedy very fast. By the end of 1980 I had twenty-one pairs, by the summer of 1981, thirty-nine pairs. That was the summer I busted my butt testing twins."

As those initial months of recruiting and studying twins spread out into one year, then two, no papers on results were forthcoming, but many stories reached the press about bizarre similarities. The Jim twins were by no means the only ones with newsworthy parallels. Boggled by the unexpected discoveries, Bouchard and his staff could not remain silent about them. While understandable, this turned out to be a strategic mistake with lingering repercussions. The defenders of environmentalism, aware the Minnesota project had a strong potential for demolishing their belief systems, were alert for a means of discrediting it. The publicity about the weird twin "coincidences" provided one.

Most of the planet's five to six billion inhabitants are quite different—recognizably different to everyone who knows them. To have two individuals violate this fundamental fact of nature becomes the genetic equivalent of two women wearing the same dress to a party—titillating and slightly embarrassing, the stuff of giggles. As the bizarre similarities of Bouchard's twins leaked out, derisive comments began appearing in the scientific press. Because no information was released about the meticulous testing, the study's core effort, outside observers knew only about the gee-whiz oddities that, lacking context, cast a pall of bogus science over the project. A report of two sisters arriving in Minneapolis wearing seven rings each became an easy-to-ridicule whopper.

A young scientist named Arlen Price, who began his academic

career as a psychologist and eventually became a prominent behavioral geneticist, was following the Minnesota study with friendly interest. Not only did he read all the press reports, he would get periodic updates from a friend on Bouchard's staff, Nancy Segal, with whom he had attended graduate school. As time went by and the amazing stories continued to appear in the press, Price telephoned Segal and urged her to persuade her colleagues to publish their substantive findings in scientific journals as soon as possible. Too much bizarre anecdotal stuff was appearing in the newspapers, he felt, and it was damaging the twin study's credibility. He reminded them of a starchy saying in their profession: The plural of anecdote is not data.

Segal replied that they could not publish until they had sufficient data for scientific conclusions; this took time. In that case, Price advised they should plug the leaks about the twin oddities. Segal assured him they were all aware of the problem.

WHILE THE SENSATIONAL Minnesota anecdotes posed a threat to the overall study, developments elsewhere in the field of psychology were enhancing the timeliness and significance of a comprehensive separated-twins study. Before the findings from the Minnesota research are examined, a brief rundown of the new trends will provide relevant context. The most important developments were two new ideas about the environment that had emerged and that were complicating the tidy nature-nurture dichotomy. These were not mere refinements on the environment concept; they were major alterations that cast doubt on all previous attempts to sort out environmental from genetic effects.

For most of the century, a rearing environment was regarded as a relatively fixed and stable entity, the setting in which the child and his siblings grew up. In addition, the environment was seen as an influence on a child's development that was separate and distinct from innate influences—nature here, nurture over there. In analyzing the determinants of personality development, psychologists examined the environment as one phenomenon, generally the family home for young children, and, more recently, considered as well the genetic inheritance. They were seen as two diverse forces operating on each child. While adding genes to the equation was a sizable advance over the years of ignoring them altogether, renovations on the paradigm were by no means completed.

Psychologists, even those who acknowledged genetic influences, were bothered by a fundamental question. If environment was so powerful in determining personality and behavior, why were there such enormous differences between brothers and sisters who had been raised in the same homes and who shared roughly 50 percent of their genes? To answer this question, Sandra Scarr and her colleagues at the University of Virginia and Robert Plomin and others at Pennsylvania State University developed the concept of *nonshared environment*, that is, the environment particular to each child in a family. This might be a special relationship with a parent, special treatment to the last born, and so on. It could also include varying biological conditions in the womb or in early nurture. Later in development it could be different friends, perhaps different schools.

When studies were run to examine this idea, it was found that the nonshared environment played a much bigger role than the shared one. In fact, the shared environment, the one previously assumed to be the *only* environment, turned out to have a negligible effect on personality development. Brothers and sisters were rarely much alike, and, to the degree that they were, the similarities were found to be genetic, not environmental.

The other major new conceptual shift was that the fundamental environment children start with may not be free of genetic influence. Previously it was thought that if a child grew up in a family of animal lovers, surrounded by pets, that milieu was considered environment, plain and simple. As such, it would be cited as the explanation for the child's fondness for animals. With the broadening genetic perspective, however, the houseful of pets might be seen as a manifestation, at least in part, of the parents' genes, elements in the parents' fundamental nature that distinguished them from others. And if the parents' love of animals had a genetic basis, these genes could well be shared to a degree by the offspring. Pets and genes were hopelessly mixed. And while the traditional psychological view had been that the pets were a molding environmental influence, it now turned out that the pets and the child's love of pets might both be expressions of genes shared by parent and child.

Ethologists had long known that many organisms go a long way toward creating their own environment—everything from earthworms producing the soil they move through to beavers living in lakes they themselves created. Their environments were seen by some ethologists as functions of their genetic makeup—or to use Richard Dawkins's phrase, such self-made environments were manifestations of "the extended phenotype." According to this concept, the line between an organism and its environment grows indistinct. To Dawkins, the beavers' dams were extended manifestations of their genes. Carried further, this idea would suggest that an individual's hobby, career, and marriage might be seen as extensions of his or her genotype. They would be colored and shaped by the environment, to be sure, as hair color and skin texture are, but manifestations of genes all the same.

An example of molding one's environment could even be seen in a child of a few months, one young enough for the environment to have little opportunity to twist and form. Except for those that cling to the blank-slate notion, most would agree that an infant, in the first months after birth, is pretty much a raw genetic package. If genes produced a fussy, irritable baby boy, he would confront very different parents than his well-behaved brothers and sisters. He would have made his environment different than theirs. In school, the hyperactive, inattentive child will find himself facing a hectoring, hostile teacher who is nice as pie to the child's model classmates. Carrying this mechanism to an extreme, the most problematic, troublemaking child may, as he grows, find himself in a special school, a reformatory, or a juvenile prison, in which case he will have succeeded in changing his environment totally.

These new ideas were major changes in the neat nature-nurture model and went a long way in discrediting the numerous studies that for years had examined and measured the environment in order to appraise the *external* effects on developing children. Genes and environment were now recognized to be deeply entwined. Each child was seen as having a unique genetic makeup and a unique environment as well, a portion of which was determined by the child's genes.

This was true of all children—except identical twins, for whom the environments varied but the genes were the same. Even though the rearing settings of separated twins might have been similar—as the critics, in some desperation, never tired of pointing out—the environments were all *by definition* different to some degree, certainly more different than the one-home environment of reared-together twins. And they would also be experiencing at least as much variation of non-shared environment as reared-together twins, probably a lot more. For these reasons, and taking into account the revised view of environment, reared-apart identical twins still provided the best conceivable study subjects for sorting out genetic from environmental effects.

Other developments made the time propitious for a definitive reared-apart twin study. Not only Wilson's *Sociobiology* but a number of other intellectual developments were leading to a growing receptivity to genetic-biological influences on human behavior and a concurrent (but often independent) weakening of the longtime conviction about the environment's overriding power to mold children. These shifts in enlightened thinking were the culmination of currents stirring in the scientific world for nearly a century and will be explored in chapters 12 and 13. After periods of exile, the biological approach to behavior was reasserting itself in the late 1970s and set the stage for the definitive research project Bouchard was undertaking.

FOUR

COSMIC SECRETS OF TWINS

To FUND THE TWIN STUDY, Bouchard had continued to receive grants from his own university; from an original backer, the Spencer Foundation; and from the National Science Foundation. These sources were limited, however, and his need for money was always pressing. Bringing twins to Minneapolis from different parts of the world and housing them for a week cost more than feeding colonies of fruit flies.

An early source of support was the Pioneer Foundation, a New York–based charitable trust established in 1937 by a textile millionaire, Wickcliffe Draper, to fund research that would "improve the character of the American people." A strong obstacle to this goal, in Draper's view, was mixing the races, and this unsavory slant came to the public's attention when the foundation helped fund the genetic research of William Shockley, the Nobel Prize winner with openly racist views.

Bouchard is unapologetic about accepting money from Pioneer. "They give grants for controversial stuff like twin studies and behavioral-genetic studies," he said. "They fund stuff on race. They also fund Phil Rushton, a very controversial character who does work on race differences. They've been accused of all kinds of things, being racist and so on. I don't think it's justified, but I couldn't care less. My rule is that if they don't make any restrictions on me—what I think, what I write, what I do, I'll accept their money. I submit my grant proposals to them, the proposals are reviewed, the Pioneer people choose to fund them. *Why* do they choose to? That's their business. Then I'm free to do whatever I want. I don't care who they are."

The head of the Pioneer Foundation, Harry Weyher, would not comment on how they choose projects to fund, on the grounds that such statements might appear an effort to influence research. Taking a worst-case supposition, that the foundation still fosters the racial, eugenicist views of its founder, they probably operate on the conviction that genetic differences between the races do exist; therefore *any* research into genetic influence on personality and behavior will advance this assumption. To believe that Bouchard would publish *only* those conclusions that reinforced whatever racial ideas Pioneer might have, or in any way to alter the direction of his research, is to believe that the Minnesota team of eighteen scientists, working with graduate students and scores of twins and in the full glare of a major university, could conspire to tailor results to the suspect beliefs of the Pioneer Foundation, just one of several money sources.

Another backer of the Minnesota studies was the Koch Foundation. In describing how this came about, Bouchard said, "I got a letter one day from a man named David Koch. I'd never heard of him. The letterhead was something like the Koch Technical Group. He wrote that he'd read a little about the study of twins reared apart by Sir Cyril Burt and that the study had been exposed as a fraud. He said he was curious about it and could I give him any information? I went through my files and pulled out a bunch of clippings and wrote him a single-page, single-spaced letter outlining what I knew about the Burt business. I ended by saying if he wanted any more information to let me know.

"A month later I got a letter thanking me, saying it was exactly what he'd wanted. Then he wrote, 'By the way, I have a small foundation. We occasionally fund research projects. Why not submit a proposal?' I ran to the research office and asked them what is the Koch Foundation. They looked it up and said it generally provides educational scholarships for offspring of their company's employees, mostly small grants of three or four thousand dollars. I thought that it didn't make sense for me to pursue a grant unless I got around fifty thousand dollars. Not worth the time. I wrote a proposal and asked for eighty thousand dollars. Two weeks later I had a check in the mail for eighty thousand dollars!"

Bouchard roared at the recollection. Later he learned that David Koch was the billionaire scion of an Oklahoman oil and gas dynasty, a still-young businessman with scientific interests who had previously supported the anthropological work on man's African origins of Donald Johansen, the discoverer in 1974 of "Lucy." Looking back on the Koch windfall, Bouchard said, "This country is great for having so many different sources of money. It's not like most countries, where everything comes from the government. Here, if the government turns you down, you can go elsewhere."

FINALLY, IN 1984 Bouchard's twin study produced a number of papers on their findings, although the researchers chose to bury their most significant finding, a concordance of .73 on I.Q., in the back of a paper about something else. All the research documented significant heritabilities for the traits measured. Also in that year, for a book entitled *Individuality and Determinism*, which was edited by Sydney W. Fox, Bouchard contributed a chapter that was primarily a defense of previous twin I.Q. studies. Here again he downplayed his own findings and instead addressed the criticisms of the earlier studies, systematically refuting each of the arguments against them in what appeared to be a warning to the critics-in-waiting not to use on him the exhausted, discredited arguments with which they had flailed earlier studies.

He takes on all the critics of I.Q. studies, paying particular attention to the most extreme, Leon Kamin, who wrote in his 1974 book, *The Science and Politics of I.Q.*: "Whatever the 'experts' may say, there is no compelling evidence that the heritability of I.Q. is 80 percent, or 50 percent, or 20 percent. There are not even adequate grounds for dismissing the hypothesis that the heritability of I.Q. is zero. The evidence is clearly inconsistent with a high heritability."

In rebuttal, Bouchard first stated the most persuasive evidence turned up by his study: the mean correlation of I.Q. between reared-apart twins of .72, as compared with .52 for *fraternal twins reared together*. (His numbers changed slightly as the number of pairs tested increased.) As might be expected, identical twins reared together had an even higher correlation— .86. Although this MZT figure is considerably larger (.14) than the MZA figure, it is not as large as the difference between the mean correlations of MZAs and DZTs (.20). This said very clearly that identical twins reared in *different* homes are closer in I.Q. than fraternal twins reared in the *same* home.

Bouchard goes to great pains to methodically address each one of Kamin's stabs at discrediting twin data. Having addressed his complaints about fraternal-identical comparisons, Bouchard went on to dispute the criticisms of reared-apart twin studies, most notably those in the 1981 book by Susan Farber, and painstakingly answers the points of each, one by one.

In his rebuttals, Bouchard did not cite one way of looking at his statistics that make nonsense of the critics' most frequent complaint: that separated twins are usually adopted into very similar homes, thereby homogenizing their test performances. In all his measures, the reared-*apart* identical twins measured consistently more concordant than reared-*together* fraternal twins. This makes it unavoidable that, regardless of the similarity of the two twins' home environments, they could not be as similar as the *one* home that reared-together fraternal twins shared.

My brilliant editor, Victoria Wilson, on reading this argument, pointed out that fraternal twins growing up together may *strive* to be dissimilar. Subtle as this point is—and it is the sort of skewing possibility for which researchers must constantly be alert—it is overcome by the impossibility of *willing* one's I. Q. higher or lower or the unlikelihood of making oneself shy because of an aggressive twin, excitable because of a phlegmatic one, and so on. Results showed, however, that traits amenable to conscious choice, like the last two mentioned, measured no more concordant than those, like I.Q., that are not. Even though Bouchard's paper is about I.Q., he once again unobtrusively works his own findings into his discussion of all I.Q. studies instead of trumpeting what many considered the definitive data on the controversial subject.

Summarizing his other findings, Bouchard writes: "In the domain of personality, estimates of heritability are somewhat lower [than in I.Q. and physiological traits] but are still substantial, somewhere in the area of .50 to .60." While he admits his concordances are somewhat higher than those of earlier efforts to measure the heritability of personality by adoption studies, he says, "both the twin studies and the adoption studies do, however, converge on the surprising finding that common family environmental influences play only a minor role in the determination of personality." This was quite a gauntlet to toss down at the heavyweights of psychological thought like John B. Watson, B. F. Skinner, and Dr. Spock, for whom family environment was pretty much everything. Even in Freud's constructs, family influences were far from "minor."

IN 1986 the Minnesota study published a paper on homosexuality in the *British Journal of Psychiatry*. The paper began by reviewing six earlier studies of homosexuality that compared MZs with DZs and that consistently found far higher concordances among the identicals, in one case 100 percent. There were, the Minnesota authors allowed, potential flaws in each of these studies. In all of the six pairs studied by the Minnesota group, two male pairs and four female, at least one twin had a history of homosexuality. Of all the pairs, one male pair was definitely concordant in that both twins had been actively homosexual well before reuniting. The other male pair was less clear and illustrated the pitfalls of self-reporting.

One brother said he was fully homosexual; the other considered himself heterosexual, but when he gave his sexual history, however, he said that between the ages of fifteen and eighteen he had had an affair with an older man. Now in his mid-thirties, he said that he is happily married to a woman but added that they made love infrequently. In the eyes of skeptics, this history did not qualify him as an award-winning heterosexual. And if his heterosexuality was disallowed, the male concordance in this tiny sample would have been 100 percent.

With the female twins, the homosexuality paper was on only slightly firmer statistical ground in that it included four pairs. In three pairs, one twin was leading an active lesbian life. A twin in the fourth pair had had a long and intense affair with another woman but was currently involved with men, so the Minnesota team designated her bisexual. The identical twin sisters of all four of the homosexually active women were exclusively heterosexual. Bouchard put this discordance forward as strong evidence that lesbianism is not genetic, that it appears to be environmentally determined. While unlikely, a behavioral trait could be genetic in one sex and not in the other; it is not impossible. Although this was just one of many traits, it is one that could spell the difference between a happy life and a difficult one, so with Bouchard's findings, the concerned, involved parent was still in business.

Several things are unusual about this paper. For most of the six years the Minnesota Twin Study had been in existence, Bouchard and his colleagues had emphasized in their papers the dangers of drawing conclusions from small samples. So it is surprising that they were so quick to draw conclusions about male homosexuality ("it is hard to deny genetic factors") based on only two sets of male MZAs. In their previous papers, the Minnesotans had been cautious about drawing *any* conclusions, even when their data was weighted toward a particular inference. Now, with a small sample in which the orientation of one subject was unclear, they were speculating broadly on the causes of lesbianism.

With a study that pointed to the *opposite* result—no genetic component to a trait—and based on a sample that was unusually meager, they almost trumpeted their shaky conclusions. The lesbian paper's final sentence reads: "If this remains a constant finding, it will be the strongest evidence known to us which attributes a major behavioral complex exclusively to environmental factors." Researchers will never admit to anything but the purest scientific considerations in the presentation of their data, but my suspicion is that the lopsided lesbian paper reveals a feeling of *relief* in Minneapolis that genes turned out to be less than all-powerful. See, it seemed to be saying, we are not as genes-giddy as you thought.

If relief was the feeling, it turned out to be unjustified. The far more comprehensive study of homosexuality done in 1993 at Northwestern University by Michael Bailey and Richard Pillard confirmed Bouchard's conclusions about male homosexuality—although with a far lower, but still significant, concordance between twenty-three MZA pairs of .52. But a follow-up study on female homosexuals contradicted Bouchard's four pairs when .48 of thirty-five pairs of identical female twins were both gay, almost half, as opposed to .16 of eighteen pairs of fraternal twins and .14 of nontwin sisters. So it appears the Minnesota team was overly eager in announcing the discovery of a major trait in which genes played no part.

FIVE

MINNESOTA'S TRIUMPHS

LATE IN 1986, the Minnesota group had completed assembling its data for a major paper, the concordances of a broad range of personality traits between sets of separated twins, now totaling forty-four. The paper also included data on 331 reared-together twins, 217 identical and 114 fraternal. A trait that showed one of the highest levels of heritability, .60, was traditionalism, or a willingness to yield to authority. The only one of these traits to show a higher heritability was social potency, which included such characteristics as assertiveness, drive for leadership, and a taste for attention. A surprisingly high genetic component—a .55 concordance—was found in the ability to be enthralled by an esthetic experience such as listening to a symphonic concert.

Good news for the environmentalists was that one important trait, one of the species' most appealing, appeared to have little genetic basis: the need for social intimacy and loving relationships. The heritability figure for this trait was .29, putting it just under Bouchard's somewhat arbitrary threshold of genetic significance. If you were warm and loving, you probably came from a warm and loving home. But here it should be remembered that if a trait has a high heritability, say above .60, that still leaves room for many people who came by the trait not by their genes but by environmental influences.

Other noteworthy numbers emerged. In five of the categories

tested—stress reaction, aggression, control, traditionalism, and absorption—the separated twins tested slightly *more* concordant than the identical twins reared together. As this oddity is not mentioned in the paper, the Minnesota team appeared to give it little significance. No trait showed zero heritability, and most were in the .45 to .60 range.

In this comprehensive paper on personality, the Minnesota team could not resist taking shots at earlier psychological studies that measured similarities in twins reared together and showed the twins to be far more similar to each other than they were to nontwin siblings. Oblivious to the possibility of genetic influence, the studies attributed those similarities to environmental factors such as twins' influences on each other, more similar treatment by others, and greater expectations from families of similar behavior from twins than from brothers and sisters. All of these explanations were wiped out when the twins studied were raised separately.

It is understandable that Bouchard, who was usually diplomatic and nonconfrontational in putting forth his orthodoxy-shattering conclusions, would sound a note of exasperation about these gene-blind studies. He knew that the researchers behind them, all trained psychologists who knew of the high level of genes shared by twins (50 percent or 100 percent), were so grounded in the prevailing dogma of their profession that they never thought to look outside the environment for explanations of why twins were more similar in personality than other siblings. Bouchard made clear that he found this an astounding misassumption. By ignoring the possibility of genetic influences on personality in setting up these studies, by placing genes off-limits, psychology's policy-makers assured none would be found.

Minnesota's paper on personality traits repeatedly reproached earlier studies aimed exclusively at the environment, saying that because they focus on "social class, child-rearing patterns and other common-environment characteristics of intact biological families, [these studies] cannot be decisive because they confound genetic and environmental factors." The phrase *cannot be decisive* was a polite way of saying they tell us nothing. It was as though scientists undertook studies of the makeup of water with the proviso that only hydrogen be looked at; oxygen was off bounds.

The personality paper concludes with a subdued summation of the Minnesota team's startling findings: "It seems reasonable, therefore, to conclude that personality differences are more influenced by genetic diversity than by environmental diversity." The paper also asserted that shared family environment turned out to have a negligible effect on personality. Social closeness was the only trait that appeared to be mainly a product of family environment.

WHEN THIS IMPORTANT PAPER was concluded, it was sent out for review prior to publication in the *Journal of Personality and Social Psychology*. The review process by other, unrelated experts in the field is a standard procedure for scientific papers and in complex studies like Minnesota's about personality measurements can take up to a year to complete. Participants in the review process often discuss their findings with colleagues at other universities; only in unusual situations is tight secrecy maintained. Still, it is rare for the results to reach the press until the paper is reviewed and published.

A prolonged incubation period prior to publication was not to be the case with Minnesota's findings on personality. The results leaked out and were proclaimed in the lead story of the "Science Times" section of the *New York Times* on December 2, 1986, a full year before the paper itself was published. The article ran with the headline MAJOR PERSONALITY STUDY FINDS THAT TRAITS ARE MOSTLY INHERITED. Daniel Goleman, who later wrote *Emotional Intelligence*, started his article with a straightforward declaration of the findings' principal import: "The genetic makeup of a child is a stronger influence on personality than child rearing, according to the first study to examine identical twins reared in separate families."

Goleman referred to 350 pairs of twins studied by the Minnesota group and later added that, of these, forty-four pairs were identical twins raised in different homes. He described the study's design and gave samples of the questions aimed at measuring personality ("True or False? When I work with others I like to take charge"). Early in the piece, Goleman stated the overall results: "For most of the traits measured, more than half the variability was found to be due to heredity, leaving less than half determined by the influence of parents, home environment and other experiences in life."

In a box running beside the article's first column, Goleman listed the eleven traits examined and gave the heritability figure for each. The highest were social potency (.61) and traditionalism (.60). Five were between .50 and .60: stress reaction, absorption, alienation, well-being, and harm avoidance. Three were between .40 and .50: aggression, achievement, and control. The last, social closeness, just squeaked above the significance level with a revised figure of .33. A pair of warm, loving twins must have arrived in Minneapolis to boost the earlier figure.

The article included quotes from other experts that underscored the significance of the findings. Jerome Kagan, the distinguished Harvard psychologist, was quoted as saying, "If in fact twins reared apart are that similar, this study is extremely important for understanding how personality is shaped." This was a generous comment since Kagan himself, with thirty years of developmental research behind him, had been a leader in the struggle to understand how personality was shaped.

MOST OF THE PAPERS that flowed from the Minnesota Twin project in the late eighties appeared in professional journals of the psychology field. While these periodicals were respected, the time had come to break out of the hermetic world of psychology and put forth a broad summary of the Minnesota findings in a publication that covered important advances in all the sciences. In the United States, the premier venue for such reports was *Science* magazine. Bouchard had met the editor, Daniel Koshland, at a conference in 1984, when Koshland was a biochemist at Berkeley. They had had a conversation about the Minnesota project and Bouchard had obliged Koshland's request to be kept informed about the twin research. In 1989, four years after becoming editor of *Science*, Koshland approached Bouchard about writing a paper that would summarize his results to date. Bouchard was delighted.

The paper, published in 1990, shows none of the reticence of the earlier papers and starts off with a trumpet blast about the finding most likely to ignite controversy, the high heritability of I.Q. The figure was slightly lower than the earlier one, .70, but still high enough to alarm the antigenes holdouts. The paper went on to summarize Minnesota's other findings, concluding with a body blow to the environmentalists:

On a number of measures of personality, temperament, interest, and attitudes, the twins reared apart measured about the same as twins reared together.

Having set forth the explosive implication that rearing environments apparently made little difference in personality formation, the authors moved quickly to soften their claims. Within the paper's initial abstract, they suggest an explanation sure to be more palatable to traditional environmentalists. "It is a plausible hypothesis that genetic differences affect psychological differences largely indirectly, by influencing the effective environment of the developing child." This was a puzzling disclaimer since the entire study was repeatedly showing the weakness of the environment in development compared with genetic endowment and might be seen as a strained effort to assure the hardliners that their cherished environment was still important (if only to transmit genetic effects).

In pointing out that their heritability figure for intelligence of .70 was somewhat higher than the earlier studies, the paper saw an explanation in that those studies primarily involved adolescents, whereas the Minnesota twins were closer to middle-aged. Other research had shown that heritability of most traits *increases* with age—that is, twins grow more alike as they grow older (a surprising statistic in light of the increased opportunity for environmental factors to work their effects)—so that the Minnesota I.Q. finding was not really inconsistent with prior studies.

As always with such papers, the Minnesota report in *Science* naturally went into detail about the study's methodology and offered *mean* figures on the degree of separation and age at reunion, among others. Because of the informed-consent agreement, which promised the twins anonymity, the article provided no biographical information about individual twins, either anecdotal similarities or specifics about time of separation or age at reunion, or any information that might be used to identify a particular pair.

The core of the paper was a table that listed all of the traits examined with the percentages of correlation of the MZAs and the MZTs. The MZA correlation figures ranged from .96 for finger-ridge count to a .33 mean for social closeness. In other words, of the twenty-eight categories listed, including both physical and personality characteristics, every figure showed at least some degree of genetic influence. Most of the figures clustered around the more than merely significant 50 percent area.

In their conclusions Bouchard and his colleagues stated that the evidence indicated that parents might be able to increase the rate at which their children develop cognitive skills, but they will have "relatively little influence on the ultimate level attained." Later, more diplomatically, they said: "The remarkable similarity in MZA twins in social attitudes (for example, traditionalism and religiosity) does not show that parents cannot influence those traits but simply that this does not tend to happen in most families."

With that caveat Bouchard held out a less frightening alternative than powerful genes: ineffectual parents. If children do not turn out as we would wish, we would rather hear that the fault lies with their parents, whom we can improve, than with the genes about which we can do little. It is with asides like this that the Minnesota group revealed its awareness of the affront their findings would be to the cadres of psychologists who had based their therapies and problem-solving strategies on environmental assumptions this study was now demolishing.

In the paper's principal conclusion, however, that in every trait investigated genes prove to be an important source of variation, they sounded a strident note: "This fact need no longer be subject to debate; rather it's time instead to consider its implications."

The paper ruminated about an evolutionary explanation for the high degree of variation among humans, the apparent fact that newborn babies are already different from each other, before the environment has had a chance to work its influences. The paper cited one theory that says there is no point to these individual differences and considers them "evolutionary debris." Another believes them to have an adaptive function that has been selected for. Many Darwinians consider the small differences the Minnesotans were measuring, the variations that make humans more interesting than fruit flies, to be the engine that drives evolution. This is hard to digest in that it suggests that if you and your sister are very different, one of you may be launching a new direction for the entire species.

The Minnesota writers left it to the evolutionary theorists to make sense of the human-difference phenomenon but said that a species that was genetically uniform would have created a very different society from ours; whatever the origin of human variation, it "is now a salient and essential feature of the human condition." Not to mention the basis of every novel, play, poem, opera, and sitcom ever written.

Behind all writings on genetic influence over personality and behavior, there are inevitably political undercurrents, as there are with any theories that touch all humans. In scientific papers these political substrata usually push close to the surface in the broad generalizations at the end that often address implications. Saying that present-day society is a product of human variation might seem safe enough on the face of it, but the assertion is surely a red flag to some who might see a hint that our social structures are, thanks to genes, inevitable. If this implication was found, it would be branded the status-quo justifying that evolutionists have been accused of since Darwin, sometimes with good reason. The paper's conclusion was remarkably simple: People are different, and they are different to a large degree because of genetic differences. While this would not seem to be a belligerent manifesto for revising political thought, to some it was just that and they marshaled their forces to combat it.

INITIALLY, HOWEVER, the response to the *Science* article was muted, with the magazine publishing only two critical letters. One was from an M.I.T. mathematics professor, Richard Dudley, who objected to the assumption of randomly varying environments for the MZAs, noting that in earlier reared-apart twin studies "some had even gone to the same school." He felt, therefore, that without precise measures of the rearing environments, generalizations about genetic influences on I.Q. to the populations at large could not be made.

It seemed odd that a professor at a university who presumably faced numbers of students for long enough periods to become acquainted with them to a degree (or that anyone, for that matter, who has ever been in a class with a group of students, as everyone has) would not accept as self-evident that within any environment—family, classroom, workplace—individuals will reveal differences in intelligence that remain more or less constant, that are not brought closer together by the educational process. But Dudley was suggesting that the high I.Q. correlations between Bouchard's separated twins might be explained by their attending similar—perhaps (good grief) the *same*—school.

Critics of twin studies who used this argument were clearly trying to leave the door open, not for improvement in the least gifted kids in an M.I.T. class, but for improvement for the least gifted kids at the lower end of American society. Underlying the point made by Dudley and many others, that I.Q. can be raised by the right environment, is a resistance to the existence of such a condition as "smart" or "stupid"; there are only good or bad environments. Although Dudley had not explicitly mentioned the underclass, Bouchard and his colleagues answered him as if he had. They quoted the warning in their paper about generalizing too broadly from their research:

> Since only a few of the MZA twins were reared in real poverty or by illiterate parents and none were retarded, this heritability estimate [of I.Q.] should not be extrapolated to the extremes of environmental disadvantage still encountered in society. . . . These findings do not imply that traits like I.Q. cannot be enhanced. . . . The present findings, therefore, do not define nor limit what might be conceivably achieved in an optimal environment.

As if this weren't sufficient obeisance to the environmentalists, the Minnesota team went on to state flat-footedly: "There is little doubt that I.Q. is malleable."

But the Minnesota I.Q. heritability figures—.75 for MZAs, .70 for all twins—were there for everyone to see. For all their concessions to those who still insisted that environment played an important role in I.Q., the Minnesota subtext was that, yes, environment can affect I.Q., but most of the time it doesn't.

In a later paper Bouchard made a point about genes and I.Q. that had been more fully developed by Harvard's Richard J. Herrnstein, who later coauthored *The Bell Curve*, and before him, by psychologist Gerald Hirsch—a point that had caused furious protest. As society draws closer to the nation's egalitarian ideal, they said, in which schools become of uniform quality, home environments are all healthy and substantial, and everyone has a just share of the nation's bounty, the genetic effects on young people's development will *increase*, rather than decrease. This would be because the wide range of environments now existing will be leveled, homogenized, equalized—leaving genes the only variable. Some might argue, and the Minnesota findings implied, that for large portions of the population, this had already happened.

The other letter raised the familiar criticism that the random environments of the study's separated twins were broad enough to show effects on gene expression. Yes, the critics said, the environments were different, but not different enough to show the environment's power over genes. The opponents would point to environmentally induced alterations in gene expression of other species, which clearly demonstrated that gene expression can be enhanced or diminished by varying environments.

The Minnesota geneticists responded that the evidence showed that, in general, genes acted "independently of the environment" as long as the environment fell within *a normal range*—that is, no droughts or toxic fumes. Vague and unscientific as the concept of "normal range of environment" sounded, there appeared to be no other way to express a condition crucial to the entire study of genetics and that would be the crux of the later dispute triggered by *The Bell Curve* in which critics disagreed heatedly with the authors over the point at which an environment becomes "abnormal" and interfered with development.

With humans the assumption is that a certain minimal level of shelter, diet, and nurturing places the developing child within the normal range of environment, the range in which the genome would express itself as nature intended. On the other hand, a child who has been beaten, starved, or chained in a closet clearly fell outside the normal range and, developmentally speaking, all bets were off. Infant Romanian orphans warehoused in subhuman conditions and others brutalized in early life are known to have suffered permanent neurological damage. Not only their bodies but their behavior can be permanently altered. But the evidence has not shown that negative, but less-than-brutal, experience will *also* cause permanent changes, just to a lesser degree. The environment must push genes pretty far before they react with any changes whatsoever in the organism. Jerome Kagan's work with Guatemalan infants, to be discussed in chapter 8, is just one vivid demonstration of the genome's power to overcome adverse conditions.

The susceptibility of behavioral genes to environmental pressures has not been shown to be a continuum on which life experience can bring about degrees of permanent neurological change. The exceptions that result from extreme environments do no damage to the normal-range concept, which remains a potent and meaningful factor in appraising genetic expression. Most professionals in the field understand this and accept the validity of the concept when discussing gene action. For humans and many other species there is a wide range of environments that permit the genes to produce their preordained results without significant alteration. The lack of scientific precision to the words *normal range*, however, carries a sense of vagueness and provides opponents with a handy weapon for criticisms and claims of "fatal flaws"—not just in twin studies but in all of behavioral genetics. Yes, the environment can thwart genetic expression; that it rarely does is a point the critics like to overlook as they trot out their shopworn examples of environmentally altered genetic effects.

SIX

TWO DOGS NAMED TOY

WITH THE PUBLICATION of the *Science* article in October 1990, the Minnesota Twin Study was solidly on the scientific map and the environment-is-everything dogma had been dealt a mighty blow. Bouchard's twins had provided statistical evidence, compelling to all but a few environmental diehards, of a genetic component to personality and behavior. He and his research team had been looking for signs of inherited traits, and they had found them—everywhere they looked. The notion of nature *or* nurture—which more and more people were increasingly realizing was a bogus dilemma—had been changed to nature *and* nurture, or, as the Minnesota group preferred to say, nature *through* nurture.

What the Minnesota group had produced were percentages of genetic contribution that were expressed in figures of heritability. This is very specific, and to outsiders a confusing term, meaning the percentage of variation of a trait within a given population that is due to genes. It is an estimate of the genetic contribution to a particular trait for the group measured—a group average, in effect—and does not necessarily say anything about an individual in the group.* Because the sample was substantial in this case, the figures had a high degree of accuracy for groups but not for individuals. It is ironic, therefore, in a study scrupulously avoiding generalizing to individuals, some of the most interesting data that was turned up—stumbled upon, rather—concerned the amazing similarities between individuals.

It should be emphasized that a number of striking and unusual behavior forms turned up that were *not* found in both twins. This is simply evidence that MZAs and MZTs do not share *all* traits and qualities, a proposition no one has ever made, least of all the Minnesota group. Such twin inconsistencies, even in unusual traits, takes little away from the potential significance of the large numbers of tastes, habits, and quirks found in both members of a twin pair.

One of the most interesting pairs of twins to undergo the week of tests at Minnesota were two Englishwomen, Daphne Goodship and Barbara Herbert. Their mother was an unmarried Finnish exchange student who was living in England in 1939 when she gave birth to the twins. Before returning to Finland immediately after the birth, the young woman put them up for adoption. The sisters were raised by two unconnected English families and did not meet again until shortly before their fortieth birthday.

What makes their story unusual is that the two adopting families were of different social classes, answering a frequent complaint of the critics that the social-economic level of the "separate" homes are similar enough to explain the parallels. Barbara grew up the daughter of a municipal gardener, a British groundskeeper in a public park. The adoptive father of the other twin, Daphne, was a well-paid metallurgist who lavished on his daughter the comforts and privileges of an upper-middle-class rearing—private schools, cultural events, and vacation trips abroad. While such disparate childhoods would seem to augur

---

*The problem with the term lies in the stipulation *within a given population*. If the trait measured was dark hair and the group was Iraqi males, there would be no variation and the heritability would be zero. If the group measured were U.N. employees, the variation would be large; and because this is a gene-based trait, the heritability figure would be close to 100 percent. Heritability, therefore, is completely dependent on the group measured. The hateful result is that you can have a heritability of zero in a trait that is 100 percent genetic. For a nongeneticist like me, this knocks the English language senseless.

a matching set of before-and-after Liza Doolittles, it did not turn out that way. The two women were more similar than most sisters growing up in the same home.

When the genetically identical women arrived in Minneapolis, the university staff was immediately struck by a shared idiosyncrasy: They laughed constantly. No matter how innocuous an interview question or routine an observation, it would send them both into peals of laughter, laughter that at times was difficult to subdue. Just when the merriment seemed to have subsided, they would go off again. Staff members dubbed them the giggle twins. Because an exchange of glances usually preceded the eruptions, it appeared they shared some secret joke, or perhaps found the Minnesota Twin Study the funniest experience of their lives. But it was learned that both women had been laughers for years before they met, and they admitted that in their pre-reunion lives they both knew no one who laughed as much as they did.

When the twins were not cracking up, they were cheerful to the point of giddiness. They were also enthusiastic pranksters. Walking across the Minnesota campus with Bouchard, they suddenly decided to pelt him with snowballs, shrieking as they forced the distinguished scientist to shield his eyeglasses from their blitz. While not exactly a behavioral oddity, it is unlikely that many forty-year-old women would do it, even if prompted by their sister's tossing the first ball. Later in the week during an interview together—a videotaped record of the twins interacting—they pulled another stunt. They announced that they had both wanted to become opera singers. Since neither could hold a tune, they thought this a fine joke but quickly admitted their hoax. While cute, it is not everyone's humor.

In all the physiological tests—brain waves, heart, and so on—they responded almost identically to stimuli, needles jumping as with the same person being retested. Both had a slight heart murmur and the thyroid glands of each was a shade larger than normal. Both were attractive women, both five feet three inches tall; while they looked alike, they could be told apart without difficulty. Both had been born with crooked little fingers, a malformation that made piano playing and touch-typing impossible.

In spite of their different backgrounds, they became best friends and relished each other's company. In the videotaped interview, they often answered in unison or finished the other's sentences. While responding to the questions, both nervously fingered the single strand of pearls that each wore. Both gesticulated when talking, sometimes so animatedly that they actually addressed their hands with lines like, "Can't you be still?" Before their reunion, each had tried to control this habit by sitting on her hands. When either was anxious, they both had a long-standing habit of placing their left hands over one side of their faces. Another odd habit they shared was wrinkling their upper lip to push up their noses, a facial expression both women described with the same term: *squidging.* All of these anomalies were present before meeting the other.

Also on the physiological side, they had each suffered miscarriages with their first pregnancies, then went on to give birth to three children, two boys and a girl, in the same gender order. Despite their apparent insouciance, both women took motherhood very seriously and were above average in their dedication to their children. Each had dyed her graying brown hair the same shade of auburn. Perhaps the most remarkable similarity, considering the different socioeconomic and educational levels of their upbringings, was that on I.Q. measures they tested nearly identically, even on vocabulary tests, which are generally considered, even by the genetically oriented, a domain highly influenced by the rearing environment.

In the area of personality the similarities grew more surprising. Both women were very energetic. Both were unusually tight with money and had a reputation among their friends as penny-pinchers. As children, both had had bad falls down flights of stairs and the accidents had left both with weak ankles. As adults, both feared falling on stairs and always grabbed banisters. Heights in general frightened each of them.

It is unnerving to think that we all might be walking around genetically programmed to fall down a flight of stairs or step into a manhole. Fortunately, logic does not insist on such fatalism. Some duplicated occurrences between twins are undoubtedly coincidence, plain and simple. Just because shared mishaps are part of a long list of parallels, many of which may have a genetic component, doesn't mean every one of them has a genetic component. In addition, it is quite possible that some parallels were part genetic, part coincidence. Twin sisters might have accidents as children because of a genetic disposition toward clumsiness; that the accidents were the same—falling down stairs, for instance—might be pure coincidence.

At school, Daphne and Barbara each had the same two strong dislikes: math and sports. Both had been Girl Guides. Growing up, both read a lot; the novels of Alistair Maclean and Catherine Cookson were favorites of each. They were regular listeners of the high-minded BBC instead of viewers of the more lowbrow commercial television. Both had been readers of the British periodical *My Weekly* but dropped it. Both had taken lessons in ballroom dancing.

The two women drank coffee rather than the more popular tea—both taking it black, no sugar or cream. They each had a passion for chocolate and for sweet liqueurs. Blue was their favorite color. Neither had a sense of direction, and they got lost easily. Their taste in clothes was highly similar. When they arrived for their reunion at London's Kings Cross Station, both women wore a beige dress with a brown velvet jacket. They deny any communication and insist it was a coincidence. Both felt themselves shy but appeared to others as unusually outgoing. Both were happy with their lives and could see no aspect they wished were different.

A strong idiosyncrasy shared by the two was that they refused to assert opinions. In spite of wily efforts, the Minnesota examiners were unable to lure either twin into voicing a position on any topic the least controversial—Northern Ireland, the women's movement, South Africa. Even though both expressed admiration for the Queen, neither would venture an opinion on the pros and cons of the monarchy or whether it should be retained—even though in England at the time, it was difficult to find anyone without an opinion on this. Neither woman had ever voted, each feeling she did not know enough about the issues; yet both had worked as polling clerks.

The women's life stories had striking parallels. Both met their future husbands at town hall dances when they were sixteen. Each husband-to-be worked for the municipal government; both men were described as quiet and conscientious. When they married, both twins did so in the autumn, with large church weddings complete with choirs. Everything about their courtships, marriages, and childbearing was highly similar, surprising in light of the different social classes from which they had come.

Alike as the two women were, there were differences. On the physical side, Barbara, the gardener's daughter, was twenty pounds heavier than her sister. While the pattern of childbearing was the same for the first three children, Barbara stopped with three, but Daphne had two more children.

A second pair of British reared-apart twins grew up in even more widely disparate social classes than the giggle twins. The father of one girl was a lawyer, and she had been raised in an educated, refined milieu and had attended private schools. The other was a lower-class East Londoner who quit school at sixteen to take a job. According to Minnesota's David Lykken, "One had a Cockney accent, and the other spoke like the Queen." When submitted to Minnesota's battery of I.Q. tests, their scores were just one point apart.

No Minnesota identical-twin set attracted more media attention than Jack Yufe and Oskar Stohr. Their story of widely diverse upbringings—one a Trinidadian Jew who moved to Israel, the other a Hitler Youth who grew up in Nazi Germany—contains so much intense and improbable drama, it threatens to overshadow the scientific implications of their astounding similarities. Maybe it should.

The twins were born in Trinidad in 1933 to a Jewish father and a German Gentile mother. At the time of the boys' birth, the parents' marriage was breaking up; within six months the mother had returned to Germany with Oskar and an older daughter, leaving Jack with his father in Trinidad. The mother took her two children to her family's hometown on the German-Czech border. After only a few months she left to take a job in Italy, leaving Oskar and his sister in Germany to be raised by their grandmother, a very Aryan-appearing woman—small, blonde, and blue eyed—who was a devout Catholic and insisted that Oskar be brought up Catholic as well.

Throughout his infancy, Oskar knew he had a twin somewhere; he also knew his father was Jewish but was told by his mother and grandmother never to tell others. The initial reason was that Oskar's mother had remarried a German Gentile who would have reacted very badly to learn his wife had previously been married to a Jew and that his stepson was half Jewish. When the Nazis came to power, secrecy became crucial as purges began, even in small border towns. To the extent that six-year-olds think about such things, Oskar thought himself a German and a Catholic. Before the war ended he reached the age at which joining the Hitler Youth was mandatory for all boys in the town.

On the other side of the Atlantic, Jack passed his boyhood as a white Trinidadian. Like Oskar, he knew he had a twin in Germany but thought little about it. His father told Jack that because he was a Jew, he would not be obliged, as his classmates were, to go to church; likewise, he could ignore the Christian activities in school and in Trinidad. Except for sporadic visits to synagogues on high holy days, these dispensations were the extent of his religious education. He was never bar mitzvahed.

Jack was a handsome, athletic boy—a champion rower and an honored Sea Scout. He had a temper and, with his schoolmates, did not hesitate to use his fists when crossed. He admired his father, who he knew was uneducated but who he would later claim was as street-smart as anyone he had ever met. Jack was dashed, therefore, when after the war his father sent him to live with an aunt, a concentration camp survivor, in Venezuela. When Jack reached his teens, his father and aunt decided he should go to Israel, where some relatives lived on the outskirts of Tel Aviv. For a few months he stayed with the relatives and eventually wound up at a kibbutz on the Sea of Galilee.

Before examining the twins' similarities as adults, it should be noted that the abandonment of both boys during their childhood by their one remaining parent was a strong environmental element they shared, even though they were living in very different circumstances in different parts of the world. People close to Jack Yufe, for example, say they believe he holds much resentment against his father for sending him off, first to an aunt in Venezuela, then to a kibbutz in Israel—all before he was fifteen. Jack acknowledges that finding himself doing hard manual labor on a kibbutz in a strange country, he brooded about what he had done to induce his father to treat him so. Similarly, Oskar was troubled by his mother's abandoning him to his grandmother.

In 1954, when Jack turned twenty, he married a Sabra from the kibbutz and settled in Tel Aviv. About a year later, he heard from his father, who had moved to San Diego. He urged Jack to join him in California, saying he would set him up in business. When Jack and his wife were planning the trip, he decided the time had come to meet his brother. The two adult men made contact, then Jack booked a flight that would take them to Frankfurt, where Oskar, also married, was working in a factory.

The reunion was not a great success. One problem was language. Oskar and his wife spoke German exclusively, so the two couples could communicate only by speaking Yiddish. In addition to being linguistically awkward, it presented more disagreeable problems. Oskar, who was trying to be cordial and familial with his brother, made it clear that it was still of the greatest importance that his Jewishness not become known. The ostensible reason was their shared mother's husband, an anti-Semite who, even after so many years, would be unable to accept his wife's first marriage. The suspicion was in the air, however, that Oskar, too, was fearful of having this fact of his parentage known for his own reasons.

As the couples moved around Frankfurt, sightseeing and eating in restaurants, Oskar admonished them to speak softly because Germans can quickly hear the difference between German and Yiddish. Since this was ten years after the war, when Germans were professing shame at their savagery toward the Jews, Jack was offended at the request for subterfuge to deny who he was. This created a bitter undercurrent to a visit already made uncomfortable by two young men—foreigners to each other, strangers, political adversaries—living together for a week in a small house for no other reason than that they shared the same genes, a fact of only slight importance to them both.

While Jack and Oskar naturally were intrigued by their physical similarity, the visit was too uncomfortable for them to sit down together and search out the numerous traits that, many years later, they would come to realize they shared. Even so, similarities were apparent. With mild interest, the men noticed that they both were highly meticulous, both liked to drink, and both had tempers. The week finally passed, and Jack and his wife left with their minds on something more important than a self-hating relative—their new life in America.

Over the next twenty years the brothers had almost no contact—an exchange of Christmas cards each year but little more. In San Diego, Jack started a clothing business, later switching to appliances. He and his wife had two daughters but eventually split up in an angry divorce. The bitterness is significant in that his first wife, one of the few people who knew Jack before and after he met Oskar, remained hostile to her ex-husband but confirmed his account of the initial reunion with his brother. She confirmed as well the prior existence of the many tics and foibles the twins shared. Years later, when the astounding list of parallels emerged, skeptics sought to exaggerate the Frankfurt meeting as a planning session for a matched-pair spectacular the twins would stage twenty-five years later. Jack's wife provided me with disinterested verification for their story. Because she felt little but anger toward her former husband, it is hard to believe she would help him perpetrate a pointless fraud.

Remarried and with a new family, Jack saw his life as going well, but he had a lingering sense of unresolved business with his brother. In 1979, when he read about the Minnesota Twins Study, Jack saw an opportunity to develop their relationship. He wrote a letter telling his story and asked if Bouchard would approach his brother about participating in the study. To Jack's surprise, Oskar agreed. The offer of a free week in America for Oskar and his wife may have overcome any resistance he may have felt about another meeting with his Jewish brother.

Jack, now forty-five, arrived first in Minneapolis and went with Bouchard to the airport to meet the identical twin brother he had not seen in twenty-five years. When he spotted Oskar coming toward him in the terminal, he had an overpowering sense of seeing himself. Jack had expected the years to lessen their similarity, but the opposite had occurred. Not just the face, but the build, the walk, the mannerisms—everything about Oskar made Jack think he was watching a film of himself.

Even stranger, they were both wearing the same metal-rimmed sunglasses, an odd shape, round lenses with squared-off corners. They both had closely trimmed mustaches. Both men wore blue shirts, not exactly the same—one was dark blue, the other light—but both shirts had epaulets. Perhaps most remarkable of all, the two shirts had four pockets on their fronts, two on the upper chest and two below. In spite of men's shirts usually having a single vest pocket, sometimes two, both twins saw nothing unusual in their four pockets, saying they wore such shirts all the time as they needed the pockets to store the pencils, pads, and all the items both wanted with them at all times.

During their week together, the duplicated traits multiplied rapidly. They discovered they shared a habit of stringing rubber bands on their wrists. They both liked to read in restaurants, and both read magazines back to front. They fell asleep quickly when watching television. Both dipped buttered toast into their coffee. One of the most specific parallels was that each twin had a habit of flushing the toilet *before* using it. The twins still had the fierce tempers of their childhoods. In the case of Jack, who had had problems with alcohol, his temper when drinking could turn violent. Oskar, who had given up drinking, never got physically violent, but his anger could explode into loud public scenes. In addition to the tantrums, both reported having sporadic attacks of anxiety.

Aside from the corroboration of Jack's ex-wife, the earlier strain between the brothers made even less likely the possibility of a hoax. Even after a day or two in Minneapolis, Jack and Oskar remained stiff with each other, and nothing about their relationship suggested a shared taste for deceptive tricks. Nor did they take much pleasure in their similar characteristics. For instance, they did not find their same get-ups at the airport funny or odd or delightful, but seemed embarrassed by the duplication and appeared to be as stunned by the coincidence as the Minnesota people. Just as police rely on a "sense" of a witness's honesty, so do psychological researchers.

As the week in Minneapolis progressed, the tension between the two men eased and they became close. Jack grew philosophical about Oskar's discomfort with his Jewish roots. That this attitude was unchanged became apparent in Minneapolis when Oskar was unwilling to tell newspaper interviewers where he lived or worked. Painful as this was to Jack, he reasoned that not only had his brother been raised a Catholic; he now had a Catholic family of a wife and two sons. But more importantly, Oskar was brought up in a world when being a Jew was not only unacceptable, it was dangerous. When Oskar admitted having admired Hitler as a boy, Jack was able to muster understanding for this as well. "It shows," Jack said, "what a smooth propaganda machine can do to children's minds."

Many of the Minnesota twin parallels appeared spontaneously during their week's stay in Minneapolis. The impromptu emergence of these phenomena increased the difficulty of ever measuring them in a scientific manner, or indeed of ever knowing beyond doubt they were not rehearsed by twins eager to astound. To that criticism the Minnesotans respond that the revelations appeared convincingly inadvertent, as did the twins' astonishment at the discoveries. The parallels surfaced in a series of chance utterances and were met with stunned reactions that, had they been counterfeit, would have required elaborate scripting and considerable acting skills.

A good example of a serendipitous parallel occurred one evening during their Minneapolis week when Jack and Oskar, out of curiosity, decided to visit a hypnotist. Seated in an office with their wives, the hypnotist waited until the room was totally silent, then began counting backward to induce trances. Suddenly, Oskar sneezed very noisily, shattering the somber mood. With weary exasperation his wife said to the others, "He does that all the time. It's his idea of a joke."

Jack Yufe felt a chill. For years when he found himself in crowded elevators, standing silently and shoulder to shoulder with a group of strangers, he would feel an irresistible urge to fake a loud sneeze and usually did so. It amused him enormously to blast the self-imposed solemnity of random humans being mechanically hauled from one floor to another. Jack knew this was an odd quirk. That his brother had the same quirk was, for him, more than merely odd—as it was, needless to say, for Bouchard and the others at the University of Minnesota Psychology Department.

SOME OF THE STRANGE matched behaviors were discovered by twins before they arrived in Minneapolis. Other times, the parallels emerged during research interviews; then too, they might turn up in casual conversations between the twins or in reactions to shared events, like Jack and Oskar's trip to the hypnotist. No matter how hard the Minnesotans tried to hide their fascination with these phenomena, the twins inevitably came to realize that the matching behaviors made them more interesting, which raised the possibility of prevarication. The giggle twins had, in fact, collaborated on such a lie when they claimed they had both aspired to sing opera. While clearly done in fun, it was the kind of deliberate distortion, having sport with the researcher and his tedious questions, that could skew results. It was this sort of hoodwinking of nosy scientists that appears to have led Margaret Mead so far astray in Samoa (discussed in chapter 12). Researchers of humans are often at the mercy of their subjects' veracity and can mistakenly assume they take the investigation as seriously as do the investigators.

With all the bona fide scientific minds grappling with the separated twin similarities, it perhaps debases the argument to even mention one ersatz effort to explain these phenomena in a genes-free way: astrology. Because identical behaviors in people who share a birth date seem grist for the astrological mills, the astral possibility is put forward with distressing frequency, which is perhaps not surprising in a country where 46 percent of people polled in 1994 by Lou Harris for New York's American Museum of Natural History did not believe in evolution.

Fortunately the notion can be disposed of as quickly as it deserves. To be sure, identical twins are born at the same time, but so are fraternal twins. Yet identicals consistently measure closer in personality and behavior than fraternals. Whether Venus is rising, setting, or just standing still, the stellar configuration is the same for both types of twins, but they are very different in the degree of shared traits. The MZA similarities and the DZT differences could, in fact, be cited as evidence of astrology's speciousness.

In the same struggle to find nongenetic explanations to twin parallels, some point to the anecdotes about telepathic communication between twins. If twins can communicate telepathically, the suggestion is, this might consciously or unconsciously lead to both wearing the same four-pocketed shirt or both fearing the ocean. To test this possibility, Minnesota psychologist David Lykken ran a series of experiments. Like others who have conducted searches for signs of telepathy between twin pairs, Lykken could find none. Then too, it is never explained how such communicative power might lead a set of twins to check the same boxes on a personality-assessment questionnaire or, for that matter, why two twins communicating regularly over the psychic airwaves would agree to falsely claim a fear of heights or arrange to wear the same dress at their first meeting. Among nontwin females, more mundane means of communication like the telephone are usually used for the opposite purpose.

As the number of these unforeseen parallels grew with each new set of twins, Bouchard and his colleagues became aware that capricious nature had presented them with a tricky public relations problem. Their dilemma was that of a hypothetical astronomer, a serious scientist to his bones, who spends his night scanning the moons of Jupiter with a powerful new telescope and, to his horror, discovers a McDonald's arch peeking over a ridge on Io. What does he do?

Announce his discovery and hope to God someone else sees it too? Or quietly pretend he never saw it?

Judging from the Minnesota team's experience, the latter course would probably be more prudent. The press reveled in the twin oddities to the same degree scientists scoffed at them. By the time the Minnesota team realized the public relations pitfalls, however, much damage had been done, as mentioned earlier, and the opponents regaled themselves with derisive comments to reporters about the ludicrous twin similarities. Since it was too late to deny having seen the McDonald's arch, Bouchard's reaction was to put a halt to press leaks.

One example of this downplaying was Bouchard's "explanation" of the two women who emerged from their respective jet planes wearing seven rings each. This was not so astounding, Bouchard reasoned; both sisters were proud of their long slender fingers and graceful hands, so it was only natural that they would adorn their physical asset with attention-getting jewelry. His aim was clear: Who would deny a genetic basis for slender fingers and well-formed hands? Further minimizing the coincidence, Bouchard pointed out that both women had flamboyant personalities plus the affluence to indulge this characteristic with enhancements like jewelry.

Another example would be when Jack Yufe noted that his twin had exactly the same walk. Bouchard said this was not surprising in that they both had long torsos and short legs, traits that might affect a walk and traits few would deny have a genetic basis. Such sheepish rationales would become Minnesota's standard reaction to the bizarre parallels: a somewhat apologetic "broadening out" of the highly specific phenomena to generalized, easier-to-swallow genetic expressions shared by twins that might have led to coincidences or would have at least lowered the odds from the McDonald's-on-Io level. After all, the alternative was too grotesque to contemplate: the possibility of a gene for wearing seven rings, not six or five, but *seven*. What a hearty laugh the environmentalist gang at Harvard could have over that!

But with similar phenomena arriving in the Minnesota labs with alarming frequency, damage control was in the air. Bouchard and his colleagues began playing up the MZAs' *differences*. While they undoubtedly found them interesting, the emphasis was surely part of the effort to counteract the seeming preoccupation with oddball similarities. Bouchard even went so far as to state that twin differences might turn out to be more interesting than the similarities. "If one twin has schizophrenia and the other doesn't," Bouchard wrote in a paper, "the reason, if we can find it, might point to a cure."

And to be sure there were many differences among the Minnesota twins, physical ones as well as in personality. A pair of Japanese-American women, raised separately in California, were so unlike in appearance that they were assumed to be fraternal; only blood tests proved them identical. They were also quite different in personality; one was friendly and outgoing, the other quiet and reserved. One had a phobia about flying, the other had no such fear. At the same time there were marked similarities, mostly physiological, but one right up there in the strangeness big leagues: They both had a cracked toenail on the same toe. Also, both had had miscarriages and both suffered from the same intestinal ailment.

Oskar Stohr suffered from narcolepsy, while his twin, Jack Yufe, did not. A number of twins had been bed-wetters as children, while their twin had no such history. In five pairs, one twin had neurotic symptoms, such as depression, that had been strong enough to require medication. The other twin was symptom free, or the symptoms were significantly weaker. It was not unusual to find one twin with a phobia the other did not share. One of two male twins in their twenties had a strong dislike for smoking; his brother smoked regularly. With the generally high heritability trait of I.Q., three pairs measured close to twenty points apart. Since the identical twins had exactly the same DNA, the many differences between them, including physiological ones, had to be environmental. The environment appeared to be alive and well.

That there were differences between MZAs merely says that identical twins reared apart or reared together do not share all traits and qualities, a hypothesis advanced by no one, certainly no one in the Minnesota group. Such environmentally induced inconsistencies take little away from the potential genetic significance of the identical habits and quirks found in both twins. In addition, pronounced differences in MZAs were the exception, not the rule, in Minnesota. Of the twins discordant for various traits, most shared a longer list of similarities. While many of the reared-together *fraternal* twins shared no noteworthy traits, of the seventy-odd sets of MZAs examined by the early 1990s, over half had at least three such odd parallels, a fact that defied probability laws of coincidence.

But it is not a matter of keeping score. The mirrored behaviors were never presented as scientific data and only leaked out because they were so surprising. For observers with no stake in the genes-environment sweepstakes, the parallels demonstrated that the genome, in certain situations, seems able to manifest itself in remarkably specific ways. To some, a number of the bizarre matches could stand by themselves as challenges to existing thinking on the limits of genetic governance.

The utility of the identical behaviors in MZAs was not just their power to broaden thinking about gene penetration. They promised more specific benefits as well. One example was the migraine headaches of the Jim twins. Not only did both men suffer from them; they did so at the same time. Their migraines had started at the same age, stopped after the same number of years, started again at the same time, then stopped for both of them—for good, it appears. According to Bouchard, "physicians had never thought such a complicated migraine pattern could be under genetic control, but now they are looking out for other signs of it."

SEVEN

MORE WEIRDNESS

SIGMUND FREUD WAS remarkably successful in persuading the world that another level of mental activity, the subconscious, was a wellspring of many of our thoughts and actions. The fake sneezes of Oskar and Jack, the tree benches of the Jim twins, and a hundred other examples of replicated behavior strongly suggest another possibility: a genetic nudge toward a particular action at a particular time. This is a concept similar to Freud's in that it speaks of a force other than pure consciousness that is guiding and influencing our decisions. The difference is that the twin parallels suggest that one source for our decisions and actions may not be a repressed childhood desire or conflict but is linked instead to an infinitesimal bit of nucleic acid in our DNA. As a theory, this is at once simpler; but in its mechanistic slant on behavior, it is far less palatable than Freud's. It is certainly less poetic.

Because the mirrored-behavior anecdotes do significant damage to existing notions of human function, they inevitably meet with strong resistance. Research hoaxes or journalistic exaggerations are a lot more digestible than alarmingly mysterious processes going on inside us all and threatening our sovereignty over ourselves. The former is an old story, the latter disturbingly new. And because these phenomena are so vivid and evoke a far narrower focus of genetic prods to behavior than previously imagined, they are resisted even more vehemently by the radical environmentalists.

Although the coincidence explanation could not be eliminated empirically, it becomes a statistical absurdity in light of the sheer number of matching traits within given sets of twin pairs. The example is often cited that in a group of twenty-four random people, there is a high probability that two will have the same birthday. The example only works to dismiss twin parallels, however, if the claim can be made that the pair sharing birthdays *also* had the same profession, the same breed of dog, the same favorite actor, and on and on.

In efforts to discredit the parallels, critics take an example in isolation and weigh the likelihood of coincidence. But this does not address the phenomena. No one suggests that twin males drinking the same brand of beer is a remarkable coincidence, but the critics still point to the popularity of that particular beer, the small number of mass-market brands, and so on. Although some of the matching behaviors by themselves are quite striking, it is the *aggregate* within a given pair that is so telling. In the case of the beer drinkers (twins who were both bachelor firemen), the next item on their similarity list was that they both drank their beer with a little finger crooked under the can. The same beer meant nothing, drinking from a can meant nothing, but the little finger under the can—and about fifteen other identical quirks this pair shared—tends to rule out coincidence.

Another effort to dismiss the phenomena argues that if you sit any two people down and ask them to look for parallels in their lives, they will inevitably come up with a few. One fact emerged from the Minnesota research that handily disproves this imaginative conjecture and makes unnecessary belaboring the large number of parallels in single twin sets. All the *fraternal* twins were invited to look for such similarities, to search their pasts, their foibles, their likes and dislikes, to find matching behaviors. Only one set was able to do so. None of the others was able to come up with one example of the sort of precisely replicated behavior that marked the separated identical twins, although fraternal twins share roughly 50 percent of their genes. It is idle to suppose that the two random individuals the critics postulate, who share none, would do better in a search for coincidences.

Future research and new methodologies will be needed to nail down the precise scientific significance of the matching-behavior phenomena; but they are there, and they have been recorded. It is now the job of the scientifically curious to push for an explanation of how the McDonald's arch came to be on a moon of Jupiter. An opportunity for exciting new knowledge could be lost by those who stubbornly insist the arch is not there simply because the apparition does not conform to preexisting ideas of the way things are.

Most scientists acknowledge how frequently new knowledge is developed by colleagues following their hunches into specific and sometimes elaborate experimental tests. If after endless experimentation, and after successful replications by others, a theory is confirmed, the success can owe as much to the inspirational hunch as to the laborious experiments. Whatever else the twin parallels may be, they provide rich material for hunches.

WITH SO MUCH outside attention being given the strange twin similarities, Bouchard repeatedly emphasized that his study was about heritability *mean figures* of concordance and not about duplicated behavior between specific twins. Even so, one of Minnesota's first reared-apart-twin papers had striking examples of these phenomena. In spite of the random selection of the study's twins, several of them suffered from psychiatric disorders. Of the afflictions, a surprisingly high number were shared by both reared-apart twins. When Bouchard and some colleagues were invited to a conference in Israel on genetic aspects of psychiatric problems, they felt that the twin pairs they had studied so far demonstrated significant correlation patterns of disturbances. They summarized their findings in a report to the conference.

The paper was based on fifteen pairs of MZAs; all except one pair had been separated within six weeks of birth. Among the similarities, two female pairs and two male pairs had speech impediments. In four of the fifteen pairs both twins had a phobia of heights, while in two other pairs just one of the twins had this fear. When both twins had a phobia, it was not always the same phobia; but in one pair it was remarkably so.

Two thirty-five-year-old British women, raised in different families and not meeting until they were in their thirties, had a strong fear of water. This presented a special problem for the two women, as their families frequently went to the beach in warm weather. On these outings both women felt obliged to join the others for ocean dips and had devised an odd way of dealing with their water phobia. Each of them managed to get their feet wet by *backing* into the sea very slowly. In both cases they would proceed in this way until the water reached their knees, at which point neither could go further. It is interesting to consider the respectful hearing most educated people would have given a psychotherapist had he or she traced the water phobia of *one* of the Englishwomen back to a childhood trauma.

Perhaps no reared-apart twin parallel was more specific than a dream that a pair of middle-aged American women both experienced years before they met. The twins also shared more routine disorders. As children, both were bed-wetters, the problem ending for both when they were twelve. As adults, but before being reunited, both had had problems with emotional stability, had abused amphetamines, and had periods of intense depression. Most striking of all, while they were still teenagers, both had had the same recurring nightmare: They felt they were suffocating because their mouths were stuffed with door-knobs, needles, and fishhooks.

While dream analysis is less popular with Freudian analysts than it once was, it would still be instructive to round up a group of the remaining specialists in this therapeutic device, tell them about the two women, then ask each to analyze the dream. Some analysts would surely sidestep the trap, but perhaps others might jump in. (The differences in the responses might in themselves prove interesting, which is one reason why psychoanalysts would never participate in such an analytic bake-off.) It would be a splendid test of the therapists' resourcefulness if, after they had presented their analyses of one woman's fishhook dream, they were asked to explain that the dreamer's twin sister, growing up separately, had the identical dream.

AMONG THE LONG LIST of matched behaviors in reared-apart twins, many of the most striking involve proper names—the two Jims naming their dogs Toy and their sons variations on James Alan; Dorothy and Bridget having daughters both named Louise; and so on. Among the twin parallels in the 1937 reared-apart twin study of Newman, Freeman, and Holzinger, one rivals any turned up by Bouchard's group. Two middle-aged male identical twins, separated for a good portion of their lives, both worked as telephone linemen and had wirehaired fox terriers named Trixie.

When attempted explanations appear too preposterous, there is a natural tendency to reject them testily and strain for more reasonable explanations. With duplications of names for children and pets, a genetic rationale is so far-fetched, critics point somewhat desperately to fashions in names that could lower the coincidence odds. Indeed, as for the breed, at the time of Newman's research, wirehaired fox terriers were one of the most popular dogs in America. If both twins had merely chosen fox terriers, the parallel would have been mild. But for each twin to affix the never-fashionable name "Trixie" to their terriers renders the fashion explanation unsatisfying.

Looking at name matchups in other twin sets, no one ever claimed a fashion in James Alan and James Allen or the two Louises of Bridget and Dorothy. But such name matches occurred repeatedly. With the high frequency among twin sets of a phenomenon for which a "rational" explanation makes little sense, I feel a degree of freewheeling theorizing is justified, if not to explain the phenomenon, at least to bring an explanation within shooting range.

In his book *Nature's Mind*, Michael Gazzaniga, who was formerly the director of the Center for Neurobiology at the University of California at Davis and who is now at Dartmouth, hypothesizes that the brain's eleven billion neurons are not just broken down into two minds, the conscious and the subconscious, but many minds, perhaps as many as a thousand modules all functioning independently. For me, this created a picture of the brain as a large office building with many companies all following their own pursuits, occasionally interacting as they do business with one another, all acting in concert only for major events like a fire or a terrorist bombing.

Brooding about the twins' selecting the same names, I pushed my metaphor further. If the office building is converted to the Pentagon, I began to see Gazzaniga's thousand minds as an army of officers, all overseeing diverse operations. At the end of a long corridor is the cubbyhole of a minor lieutenant who has only one duty: He is in charge of naming things. While the brain's panels of high-ranking brass make major decisions about logistics or battle strategies, the lieutenant sits idle, indifferent, bored, his phone silent. But when it is time to name something—a new base, a jet fighter, a radar system—this highly compartmentalized bureaucrat springs to action. That's all he does, name things when the need arises. Unfortunately, however, he is a fellow of almost no imagination. He likes a handful of names and that's it. When asked to name a missile launcher, a tank, or a base mascot, he consults his name repertoire and comes up with one of his handful of possibilities.

If somewhere there existed a clone Pentagon, Canada's for instance, that was identical in every aspect, including the naming lieutenant in his cubby hole, it is conceivable that this minor officer, presented with similar items needing names, might occasionally arrive at exactly the same name for the same object. My analogy is of course preposterous, and grievously debases Gazzaniga's splendid brain model of a thousand brains. (He cannot even be blamed for the edifice analogy.) On the other hand, wild phenomena like two terriers named Trixie call for wild theories. Science is, at this point, so far from understanding such phenomena, it might require a few wild leaps to bring us anywhere near an accurate explanation.

(It is not hard to get carried away with my brain/Pentagon metaphor. As I write this, I am sitting alone in a pleasant house in Bucks County, Pennsylvania, where I know no one. I chose the locale because it seemed a good place to write. I discovered, however, that for every other aspect of my life, the isolated house was a dubious choice. I saw an explanation in the Gazzaniga paradigm. I have a committee of, say, 50 brains that are in charge of sound career decisions like choosing optimal locations for work. They put me in unfamiliar country in a preemptive burst of career responsibility. After the move was made, however, my other 950 brains realized what had happened and cried with one voice, "What the hell are we doing isolated out here in cow country?")

With or without recourse to such whimsical flights, somewhere in Gazzaniga's brain theory may lurk the beginnings of an explanation for all the reared-apart twin parallels. If, as he postulates, our mentalities are compartmentalized into many independently operating mechanisms, it is conceivable that a number of these are more susceptible to genetic influence than others. And if it is demonstrated that genes are, in fact, involved in such specific behaviors, a number of genes would undoubtedly be involved. This would greatly reduce the chances of such phenomena occurring in fraternal twins. Just as in a game of poker, you might have in your hand the same number card that another player has in his—a three, say, or an eight—but to have all *five* of your cards numerically match all five of his would be extremely unlikely. As mentioned above, remarkable behavioral matches were, in fact, all but nonexistent among the fraternal twins tested.

In the same way, striking personality facets shared by a set of reared-apart twins rarely turned up in other family members. Pondering this, Minnesota's David Lykken worked out a theory that he called emergenesis and published it in a 1982 paper which later was his inaugural address as president of the Society for Psychophysiological Research. Lykken believed that for certain traits to appear *at all*, it required a complex combination of genetic elements, *every one of which was essential for the trait to appear.*

This was different from the well-known genetic process of epistasis, which refers to interactions of genes to produce a variation. Lykken's process was an all-or-nothing bouquet of gene interactions that might produce a complex, distinctive trait such as musical genius or great facial beauty in identical twins but would not show up *to any degree* in other relatives as they didn't have *all* the sine qua non genetic pieces. Lykken also suggests this genetic process might explain that the mathematical genius Karl Frederich Gauss could be born to a bricklayer and a peasant woman of modest intelligence and that his children showed none of his genius. It appeared that he represented a onetime configuration of genes, each one of which was essential for the brainy result, and that probabilities dictated could *only* have been duplicated in an identical twin.

Given the genetic card-shuffling during meiosis—when half the mother's genes mix with half the father's to produce the child—the chances of an offspring receiving the precise configuration that produced such a rare trait in the parent is extremely low. Likewise, chances are extremely low that if one child has the rare configuration any sibling or any other relative would have it. It might require twenty-eight genes to produce Cindy Crawford's supermodel face. Her sister might have all but one of those genes and be superplain.

Although Lykken's emergenesis theory suggests an explanation for the identical behaviors, neither he nor the others on his team were prepared to hypothesize a specific genetic basis to these phenomena and instead limited themselves to reporting them and allowing others to infer what they wished. But granting *some* genetic basis, however improbable and elusive, the twin oddities would seem to fit Lykken's model, in that there had been no reports of such odd parallels appearing in other members of the twins' families: no backward-bathing relatives, or even sideways bathers; or among the Jims' relatives, no semicircular tree benches; and for Jack and Oskar, no holiday dinners with relatives faking coughs or sneezes. These phenomena occurred only in the twins; no one else came close. Because of their uniqueness, the twin-shared traits, with their thundering specificity as well as the absence of such similarities in fraternal twins or in other family members, may well be as much a product of Lykken's emergenesis as facial beauty or mathematical genius.

EIGHT

OTHER BEHAVIORAL GENETICS STUDIES

BECAUSE OF THE EXOTIC NATURE of Bouchard's reared-apart twins, the Minnesota study received most of the spotlight the popular press allotted to behavioral genetics. The general public, in fact, had little idea what behavioral genetics was or that research in the field had been under way for two decades in universities around the United States, Europe, and Australia, all turning up solid evidence of genetic influences on personality and behavior. One reason was the lingering distaste for the idea that genes have anything to do with who and what we are. Another was the laborious, undramatic methods used by the studies. Bouchard's exotic reared-apart twins not only provided more conclusive evidence, they were in themselves more press-worthy.

Except for a large reared-apart twin study in Sweden, the other genetic studies on human behavior have been based either on comparisons of adoptees with unrelated siblings or comparisons of identical with fraternal twins—all reared in the same homes. In some studies, the adoptees were also compared with both their adoptive parents and their biological parents, producing data from which a number of genes-or-environment inferences could be drawn. Whatever the methods, all the studies were efforts at teasing apart genetic from environmental influences to arrive at estimates of the percentages of heritability for specific traits and aptitudes.

Irving Gottesman has long been considered one of the grand old men of behavioral genetics, an epithet that says more about the youth of the field than the age of Gottesman, who was born in 1930. Because his doctoral dissertation, *The Psychogenetics of Personality* (University of Minnesota), was written in 1960, when genetic theories of behavior were still anathema in most parts of the academy, this definitely establishes him not just as a pioneer but as a courageous one. Although a psychologist rather than a psychiatrist, Gottesman has had a lifelong interest in the genetic basis of schizophrenia and in the mid-1960s worked with the famous British twin researcher James Shields on schizophrenia in twins, publishing with him a number of papers and, in 1972, a book, *Schizophrenia and Genetics.*

Another prominent psychologist, Gerald McClearn, having worked for years on the genetics of animal behavior, has concentrated increasingly on human behavior in recent years. Along with J. C. Defries, he was an initiator of the behavioral genetics studies at the University of Colorado and later continued work in this field at Penn State. McClearn is also a participant in the separated-twin study at Stockholm's Karolinska Institute and is one of the designers of the MacArthur Longitudinal Twin Study (twins measured at different ages), which is supported by the John and Catherine MacArthur Foundation. In 1973 McClearn and Defries wrote *An Introduction to Behavioral Genetics,* then in 1960 the two scientists, along with Robert Plomin, wrote *Behavioral Genetics: A Primer.* Perhaps because they started out in a climate hostile to their young field, these psychologists have consistently turned out work marked by meticulous methodology and scientific caution.

In the United States, one of the most venerable behavioral genetics studies is the Louisville Twin Study started at the University of Louisville in 1963. Each year since its inception, the Louisville study recruited some two dozen reared-together twin sets to produce a sample now totaling more than five hundred pairs of both fraternal and identical twins. The study, headed in recent years by R. S. Wilson, is ambitiously longitudinal, with twins tested at three-month intervals their first year, six-month intervals until they are three, then annually thereafter. In the early years of the study, the twins were assessed for I.Q. development and physical growth, but in 1976 the focus shifted to temperament.

The study is ongoing, but results so far show correlations on mental development (the Bayley Index) of .68 for MZs and .63 for DZs, only a slight difference. Correlations for nontwin siblings, however, are far lower, about .31. With temperament, on three different traits measured (by the Bayley Infant Behavior Record test), the correlation for extroversion was .43 for MZs, versus .07 for DZs. For task orientation (a measure of concentration) the correlation was .49 for MZs, versus .23 for DZs. Both of these results showed significant genetic influence. On the third trait measured, activity (level of motion), no significant amount of genetic influence was found.

The Louisville group has also recently released findings from a series of tests based on videotape recordings of infants' behavior in laboratory situations. With only thirty pairs the results so far have been inconclusive except for one measure, a trait called behavioral inhibition. For this the MZ pairs correlated .71 as opposed to .25 for the DZs, an indication of a remarkably high degree of genetic influence. This jibed with Jerome Kagan's work at Harvard on timidity in infants, which is discussed later in this chapter.

The most comprehensive and long-ranging studies of twins and adoptees have been in progress at the University of Colorado, most especially a longitudinal twin study, one of a number of MacArthur-funded research projects concerned with childhood transitions. It is increasingly apparent to researchers that it is not simply a matter of genes' influence being present or absent with behavioral traits, but that they kick in or go dormant at different periods throughout a life. This gives importance to the longitudinal studies that measure the same twin sets periodically over a number of years.

Colorado's longitudinal study is a collaborative effort directed by Robert Ende and a group of Colorado colleagues and includes as well Jerome Kagan of Harvard, Steven Reznick of Yale, Joseph Campos of the University of California at Berkeley, Robert Plomin of Penn State and London's Institute of Psychiatry, and Carolyn Zahn-Waxler of the National Institutes of Health. That this landmark behavioral genetics study is a combined effort of the N.I.H. and five of the nation's foremost universities belies the critics' efforts to depict behavioral genetics as a fringe discipline.

Before launching this ambitious twin project, Colorado had run an important longitudinal study of adoptees that began in 1975 and was completed in 1983. It was based on 245 families with adopted children and a control group of another 245 families with nonadopted children. All the children were assessed annually, from ages one to four, in cognitive abilities, personality, interests, and talents. They were tested again at seven, nine, and sixteen years, and their performance was compared with results of similar tests administered to adoptive parents, biological parents, and unrelated siblings. This study showed a substantial genetic influence on mental development at twelve months and—because of negligible correlations between unrelated siblings in the same homes—virtually no influence from shared environment. In the measurement of temperament, the Colorado study showed genetic influence on all three of the traits measured, unlike the Louisville study, which had found it in only two.

Among the intriguing findings to be extracted from Colorado's wealth of data was that for certain traits the fourteen-month-olds showed considerable genetic influence, for other traits almost none. When the same children were measured later in life, significant heritabilities were found for *every* trait. This clearly demonstrates that behavioral genes turn on at various ages. This study and other similar ones have turned up the curious fact that when the genetic component of a trait is shown to vary with age, the variance always increases, never decreases. This is surprising in that the older a child is, the more opportunity the environment has had to work its influence. But instead of gene influence diminishing, as one would expect, it increases with age. This is bad news for those who dislike sharing personality traits with their parents. With years, the similarities will probably increase. The delayed unfolding of genes also gives new resonance to the expression "finding oneself." It may be more a matter of waiting for oneself to arrive.

In 1986 the University of Colorado initiated an even larger longitudinal study, one that included 300 pairs of reared-together twins who will be assessed at fourteen, twenty, twenty-four, and thirty-six months for cognitive abilities and an array of mood and temperament characteristics. In a preliminary 1992 paper reporting on the results at fourteen months, the data showed the identicals testing consistently more concordantly than the fraternals. Persuasive as their results are for a genetic contribution to the traits measured, they end their paper by pointing out weaknesses in their twin design, a self-attack as destructive as the usual attacks from critics on gene-behavior research. While undoubtedly included to forestall this very thing, it has an undermining effect that is puzzling.

THE THIRTY-FIVE YEARS of work in developmental psychology of Harvard's Jerome Kagan has made him a leader in his field. His main area of interest has been the inhibited and uninhibited child— primarily because this is an adult characteristic that is also observable in very young infants and can be monitored over years. His numerous writings on this and other subjects, his scientific dedication, and his cant-free style of communicating have all combined to make him one of the most honored and respected American psychologists.

Like most everyone of his generation, Kagan started out as a staunch environmentalist, a disciple of Skinner and Watson. When he decided to go into developmental psychology, it was not as an abstract quest for pure knowledge but was colored by strong altruistic ambitions. This was a period when all behavior was considered learned, including the behavior that produced social problems. If psychologists could unravel the precise mechanisms of learning and pinpoint the environmental causes for children developing as they do, educators and social scientists would be positioned to improve society greatly by averting the developmental mishaps that lead to poverty, ignorance, and crime.

Kagan's first research was in 1957 at Antioch College, when he was brought in on an existing developmental study of inhibited children who had already been assessed in the first months of life. For his initial assignment, Kagan was asked to conduct follow-up interviews with the study group, who were now young adults. Along with his colleagues, Kagan was surprised to find that the children who had been extremely fearful as infants remained so throughout adolescence and into adulthood.

He later recalled thinking how remarkable it was that the effects of the parents' rearing during the first months of life could have such an indelible effect. So imbued was he with the environmentalist creed, he, like his colleagues, never entertained the possibility of a genetic explanation. When I interviewed him in 1994, Kagan, looking back on this period, said his blinkered vision was especially remarkable since in a nearby Antioch laboratory a team investigating heart-rate variations had found a connection between fearfulness and heart rate. While this finding strongly suggested a biological basis for fearfulness and a link between physiology and psychology, the researchers took the idea no further with a kind of "we don't want to go there" reticence. The concept of fearful children having been born that way was totally alien to mainstream psychology in the 1950s. No one was born any "way"; we were all at birth tabulae rasae, innocent, raw, and ready for the world to do its worst. Such was the faith in the environment's omnipotence in the heyday of behaviorism.

For Kagan this faith began to weaken in the early seventies, when he spent a year in Guatemala observing infants in a remote mountain village. The village children were of particular interest because of an unusually deprived first year of life that sprang from the Indians' tradition. Local custom had the mothers isolating their children inside cramped dark huts for the first year, never allowing them outside, never playing with them, and almost never speaking to them. As a result, the children at the ages of one and two were observed to be unusually passive, quiet, and unresponsive. To Kagan, some appeared borderline retarded.

Setting to work, Kagan and his colleagues set up a controlled study in which the village kids were measured on a number of cognitive functions. Those results were then compared with a Guatemalan group raised in a more normal fashion near a city and with a group of middle-class American children. The three groups might roughly be termed understimulated, normally stimulated, and overstimulated.

The results showed that while the deprived village children measured lower in all tests in the first years, they tended to catch up as they grew older and their rearing setting became more similar to those of childhoods everywhere—outdoor play, interaction with parents, with other children, and so on. By eleven and twelve the Guatemalan children could be considered normally developed. This transformation was heartening to environmentalists like Kagan and to socially concerned individuals who placed hopes in intervention programs like Head Start which help underperforming children catch up with their age groups. Good behaviorist soldier that he was, Kagan wrote in an early paper on this study, "The data prove the potency of the environment." Without hesitating he credited the good environment with remedying the effects of the bad.

But the data, he came to realize, could be seen in two ways. With the Guatemalan village children, there had been no intervention, no remedial program; they had merely been delivered from the negative environment of their first year, the dark hut. Something else seemed to be causing their improvement, and future evidence indicated it was their own normal development, in all likelihood their own genetic makeup triumphing over an adverse environment. Rather than the normal "outdoor" environment curing the negative effects of the grim first year, it may have merely allowed the child's personality to unfold as its genome, or as "nature," intended. While Kagan did not express it in those terms, he would later say something very close to the same thing, but in a turned-around way: "My belief in the power of early experience [in children's development] was shaken."

Shaken, perhaps, but Kagan, like the rest of the psychological profession, continued to believe in the environment as the most important molder of personality. His epiphany came fifteen years later, when he was working in Boston on a longitudinal study of infants, observing them from seven to twenty-nine months, with the aim of assessing the effectiveness of day care. The group was made up of fifty-three Chinese-American infants and sixty-three Caucasian children. Part of the entire group had from the age of four months attended an experimental day-care center set up for the study, part had attended other day-care centers, and part had been raised at home.

In the course of the experiment, Kagan noticed something unanticipated. The Chinese children, little more than babies, whether attending day care or raised at home, were consistently more fearful and inhibited than the Caucasians. The differences were obvious. The Chinese children stayed close to their mothers and were quiet and generally apprehensive, while the Caucasians were talkative, active, and "prone to laughter." These characteristics were confirmed by the mothers as typical of their children's behavior at home as well. In addition, the researchers discovered that the Chinese tots had less variable heart rates than the Caucasians. Kagan could not avoid the clear evidence of an innate difference between the two groups of infants.

It is ironic that this scientist's conversion to a biological-genetic point of view came along the lines of racial differences. Kagan was a political liberal who only three years earlier had been one of the most vociferous critics of Arthur Jensen's theories on the heritability of I.Q., theories that he and most everyone else denounced as racist. Now he was publishing his observations of fundamental personality differences between racial groups. When we conversed in his Harvard office many years later, I asked Kagan if there had been an uproar similar to the one Jensen provoked.

He smiled. "We got no flak on the Chinese paper. All the reports of the book were about our day-care findings. Everyone ignored the fact that the Chinese children were different. I think it was because they were Asians, and Asians do well. If they had been black, we probably would have gotten flak."

Asked if it was dismaying for him, an unwavering liberal, to observe inherent racial differences, Kagan snapped, "Nature doesn't care what we want." More reflectively, he added, "I wasn't so much dismayed at my observations of the Chinese kids. . . . I was a little bit saddened to see the power of biology."

His major conclusion was stated in a 1988 paper in *Science*: "We suggest, albeit speculatively, that most of the children we call inhibited belong to a qualitatively distinct category of infants who were born with a lower threshold for limbic-hypothalamic arousal to unexpected changes in the environment or novel events that cannot be assimilated easily."

It is revealing that in a sentence that is basically saying "kids are born that way," Kagan, who had thrown himself into a self-education program of brain biochemistry, manages to avoid the word *genes* yet does find a way to work in the word *environment*. In the article's abstract he speaks of "the inherited variation in the threshold of arousal in selected limbic sites." But still, not a gene in sight. He appears to be worried about the limbic-hypothalamic arousal of the environmentalists down the hall. Or the word *gene* may have been forbidden at Harvard at the time. A more likely explanation is that Kagan, as a new convert to the biology of behavior, could not switch so quickly to the terminology of the infidels.

In his study of brain chemistry, Kagan grew convinced of the major effect on the traits he was examining of the liquid mix of neurotransmitters, hormones, and opioids that surround the operating parts of the brain. This mixture, which he refers to as "the chemical broth," is made up of 150 chemicals in combinations that vary within individuals. While he believed there might be hundreds of possible combinations, his research had found two: ones that raise the threshold of excitability leading to inhibited children and ones that lower it to produce uninhibited children. This was a remarkable area of investigation for a behavioral psychologist who for years had believed in the sovereignty of the environment and had fought off any notion of biological or genetic influence on behavior. Kagan was, so to speak, in the soup.

Today, Kagan is hard on himself about his earlier unquestioning allegiance to the day's psychological dogma. "I wince at my credulity when I told several hundred guileless undergraduates in 1954 that rejection by the mother could produce an autistic child." He also speaks of his regret at wasting so many years ignoring the genetic contribution to behavior; he forthrightly admits that the omission confounded all his research. Although, like Kagan, few in his profession still believe in the environmental determinism that dominated psychology in those years, few have Kagan's honesty in admitting the scope of their mistake.

Speaking about his changed perspective, Kagan said, "I was the classic politically liberal environmentalist who believed that genes had minimal effect on behavior. Now, I am quite a different person. My data has pushed me toward granting much more power to genetic mechanisms than I would have believed twenty years ago. I arrived here honestly, without prejudice, which is a good way." In another interview he said, "For the first twenty years of my career, I wrote essays critical of the role of biology and celebrating the role of the environment. I am now working in the opposite camp because I was dragged there by my data."

In his 1994 book with a subtly significant title, *The Nature of the Child*, Kagan speculates about ways in which the genetic differences in brain chemistry of inhibited children might have evolved into the species. Noting that the thin body builds (ectomorphs) and blue eyes that are more typical of inhibited children predominate in northern Europe, Kagan puts forward the possibility that when humans migrated from Africa and arrived in northern Europe some forty thousand years ago, evolution might have favored mutations that would be beneficial in the more challenging cold environments. His experiments had shown that inhibited behavior is linked to a greater efficiency in the sympathetic nervous system and an increased production of a major neurotransmitter, norepinephrine. Kagan wrote: "Because the metabolic steps in the manufacture of norepinephrine are mediated by several different enzymes, some of which are controlled by one or a few genes, it is possible that such a change in DNA occurred."

Not only was Kagan tracing a human behavioral facet, inhibition, to specific genes, he was also speculating about evolution's role in bestowing those genes upon certain groups of humans. Whatever weight the theory might have, for the field of behavioral genetics it was exciting to have a mind of Kagan's caliber and a psychologist of his experience to be thinking at last in genetic-evolutionary terms.

KENNETH KENDLER, a wiry, fortyish psychiatrist at the Medical College of Virginia, is another prominent researcher in behavioral genetics, or more specifically in his case, psychiatric genetics. Kendler is seeking evidence of genetic underpinnings to mental disorders—primarily anxiety, depression, and schizophrenia. In the early nineties he completed a major study on alcoholism in women. Although a generation younger than Jerome Kagan, he also came through his education at a time when environmentalism held sway. In American psychiatry this meant psychoanalysis, which focused on childhood experience, that is, the environment. It ruled out inheritance, some would argue, to a degree Freud never intended.

Kendler was aware that in the first half of the century, many distinguished European psychiatrists stood aloof from the Freudian-analytic *Anschluss* and continued to investigate biological, neurological, and genetic causes of mental illness. With no effort Kendler could name all the most prominent Europeans who kept the physiological torch alive throughout the analytic heyday—Emil Kraepelin, Ernst Rudin, Wilhelm Wundt, Alois Alzheimer ("he was looking for schizophrenia, but found something else"), Elliott Slater, Franz Joseph Kallman, Eric Intermutter, Eric Stromberg. Not only could Kendler name all these now obscure scientists, but he could also state under whom they studied, their areas of specialty, where they migrated with the neurological message, and their batting averages on advancing understanding of biological or genetic components to mental illnesses.

After doing his residency Kendler became one of a growing number of young American psychiatrists who lost faith with the analytic schools and began investigating more physiological approaches to mental illness. (A number of developments converged to bring about this change of direction in Kendler and many like him and will be examined in chapter 10.) A big part of his interest in schizophrenia stems from his belief that it causes more suffering than any other illness (primarily because of the large numbers of people affected, both the afflicted and their families), but also because it is a life sentence. Kendler estimates that in the 1990s there are between twenty and thirty major research projects in the U.S. aimed at finding the genes for mental illnesses. "That's an industry," he commented.

In Ming Tsuang's 1997 book, *Schizophrenia: The Facts*, the author summarizes the large amount of work done on schizophrenia, including his own, and concludes that while there is a genetic component, it is not as strong as most experts had believed. Although no one is yet certain of the other causal factors, the suspicion is that it is something prenatal, a viral infection in the womb, a head injury, or a faulty oxygen supply. There's no mention of the "refrigerator mother" (a leading culprit for the Freudians) or similar psychological traumas suffered during childhood.

Other projects Kendler has taken part in were a study of bulimia and a study of sleep disorders; both afflictions were found to have mild heritabilities. He also coauthored a paper that sought a connection between major depression and smoking. It concluded that while a relationship definitely could be found between the two phenomena, it was not a causal one. Instead, the conditions most probably resulted from genes that led to both behavioral manifestations. His main work of recent years has been in studying various affective disorders in women—anxiety, depression, alcoholism, and phobias. This research culminated in a barrage of papers in 1992 written with colleagues in Virginia, at the University of Michigan, and at Washington University Medical School in St. Louis.

The studies were based on female reared-together twins located through the Virginia twin registry, and the numbers of twins were large, ranging from one thousand to over two thousand. Females were chosen rather than males because they have "substantially higher prevalence rates for mood and anxiety disorders." The studies were based on personal interviews with each twin, a far costlier method for obtaining data than the often criticized self-report questionnaires. The papers that resulted from the study contained detailed accounts of the recruitment methods, definitions of the conditions, and the way in which the statistics were analyzed.

One paper gave the results of a twin study of major depression in women. Using nine different definitions of depression, 1,033 women were measured for each one. The concordance rates never went above .49, some substantially lower. But that the identical twins had consistently higher concordance rates than the fraternal twins was taken as solid evidence that "genetic factors play a dominant, but not overwhelming, role in the cause of depression." It was also determined that when the environment was a causative factor, it was an environmental element specific to the individual, not an element common to the entire family.

Similar results were reported in papers on generalized anxiety disorder, again evaluating 1,033 women and coming up with a modest heritability estimate of .30, which in itself was of borderline significance. Still, the consistently higher concordance of MZs over DZs proved that shared family environments had almost no effect; the variation came instead from environmental experiences specific to one twin. Without actually saying it, this specifying of nonshared environmental experiences as causes for depression had the effect of ruling out upbringing, to the degree that twins raised in the same home had similar upbringings. The studies might be summarized by saying that major depression and anxiety disorders do not result from upbringings, which twins share, but rather from genetic factors working in concert with life events and other environmental circumstances specific to one twin but not the other—and this might include the prenatal environment.

The Virginia group also published a paper at the same time on phobias in women that was based on over two thousand female twins. The heritability of susceptibility to phobias was found to be between .30 and .40—not high, but certainly high enough to reveal a genetic component. The study discovered, however, a significant difference in the four types of phobias investigated, to the point that one type, agoraphobia (fear of open spaces), seemed to have little relationship to the other types. It had a higher heritability and a lower susceptibility to specific environmental events. Other phobias, a fear of snakes, for example, might be activated in the person with a genetic leaning toward a phobia of animals in general but was triggered by a nasty run-in with a rattlesnake. Agoraphobia, on the other hand, appeared to be either present or not and relatively unaffected by experience or the environment, although the person fearing open space would obviously be happier living in a big-city apartment than on a North Dakota ranch.

Later in 1992, Kendler's Virginia group came out with a paper on inherited factors in alcoholism in women, based on 1,030 female twins. Using three different definitions of alcoholism that separated out levels of dependency, the Virginians found in each category a substantially higher degree of genetic influence than in other conditions Kendler examined. Those studies had turned up heritabilities in the .50 to .60 range. His paper cited a number of earlier twin studies on alcoholism, most of which dealt with men (the majority were recruited through lists of people who at one time had been hospitalized for alcoholism). Kendler's sample of women was found through a population-based twin registry. In addition, all of his subjects were interviewed face-to-face, whereas previous studies had relied on self-report questionnaires. The earlier studies with men had all indicated a meaningful genetic component, but alcoholism often manifests itself differently in men and women (age of onset, duration, and so on), so it was thought that the male genetic findings might not apply to women. Kendler's study established that they do.

Scientists like Kendler, Kagan, Gottesman, and McClearn have for years now been plugging away at broad traits and behaviors—timidity, depression, alcoholism, schizophrenia—that affect large portions of the population. Their findings were later substantiated by the Minnesota twin studies. Although Bouchard and his colleagues were using a different method, reared-apart twins, they were looking for the same things, the same influences—and finding them. The far larger numbers of reared-*together* twins needed to produce statistically valid conclusions made the earlier studies far more cumbersome and laborious. Also, lacking the juicy oddities of the separated identicals, the Kendler-style studies didn't receive the media attention that Bouchard had. But their painstaking efforts laid the groundwork for the Minnesota Twin Study and all subsequent quests for behavioral genes. They were the pioneers who proved that with many of the most common psychological problems, genes were often involved.

NINE

STARS OF THE NEW FIELD

OVER HER THIRTY-YEAR CAREER, Sandra Scarr has built a reputation as one of the country's leading psychologists, having worked at Harvard, Yale, and the universities of Pennsylvania and Virginia. Like Jerome Kagan, she is a major figure in the child-development field and has done a number of studies that relate to the politically loaded subject of day care. More recently, she has emerged as an influential theorist of behavioral genetics and one of the most formidable adversaries of the environmental determinists. Unlike most of her behavioral genetics colleagues, who maintain a scientific dignity in their debate with critics, Scarr doesn't hesitate to voice her exasperation with tactics she sees as unfair and obstructionist. Where others might characterize an opponent's view as "unsubstantiated," Scarr is more likely to call it "nonsense."

Among the ideas Scarr has championed is the concept of children molding their own environments. She also belongs to the growing group of psychologists who believe most environments have a confounding genetic component in that they are fashioned in part, if not by the individual, by biologically related humans. Either way, genes are involved. The most recent data, she feels, convincingly demonstrates that nonshared environments influence development but that the environment shared with other family members has negligible effect. This would seem to indicate that the genetically loaded environment the child fashions is of far greater importance than the genetically more neutral family setting. Beyond these refinements of her overall view of child development is the conviction that a child's genetic endowment is of overriding importance to the adult he or she will become, provided the child is raised in what she calls a normal range of environment.

Her "normal-range" concept and spin-off notions like "good-enough parents," although sounding vague and unscientific, dominate her view of child development. She believes that unless a child is subjected to environmental extremes, such as constant hunger or abusive or neglectful parents, the child will develop along lines set out by its DNA. She also believes that rearing variations within this "normal range"—such enhancements as special attention from parents, superior schooling, stimulating friends—have little effect on the final product. Yes, she says, echoing the Minnesota researchers, the environment can alter development; no, it generally doesn't. Looking at this idea from the broadest possible range, she said, "Fortunately, evolution has not left development of the human species, or any other, at the easy mercy of variations in their environments."

This idea was an affront to those who had erected entire social programs, if not political systems, on the belief that personalities and behavior were direct products of the environment, that even slight improvements in the rearing matrix will produce improvements in the developing human. (The rearing environment's negligible benefits will probably also come as bad news to the parents who have pulled strings and sacrificed to send mediocre children to Princeton and Vassar.) Because of Scarr's longtime study of day care and other intervention programs for children, her thumbs-down on such efforts had more than the usual political resonance. Her conclusions about the genome's power and the environment's lack of it, drawn from her own studies and those of others, have provoked much controversy.

Like Kagan and many psychologists educated in the heyday of environmentalism, Scarr now deplores what she calls the "banishment" of biology and genetics from the behavioral sciences for decades. She is less perturbed than other behavioral geneticists, however, by the virulent resistance of die-hard environmentalists. She admits to having a contrarian nature and says it pleases her when her scientific findings go against prevailing beliefs, calling this characteristic her "intellectual perversity."

She discovered this appetite for bucking received wisdom as an undergraduate at Vassar when environmental behaviorism ruled the day. She was particularly irked when she was told there were no biological causes for differences in I.Q., that the differences were all environmental. "I *knew* that wasn't true," she told me many years later. "It couldn't be true. Parents know it's not true. I couldn't stand it. I knew it was an absurd position. So I went to graduate school at Harvard, where the attitude was pretty much the same, but I found Irving [Gottesman], who was there at the time, and I began doing research with him." She later added to her condemnation of her chosen branch of social science in the years she entered it: "Psychology was defending positions which the public knew were untrue."

In 1964, well aware she was challenging her superiors' formulations, Scarr did a twin study to establish a genetic basis to motivation—the need for achievement. At the same time, her intellectual soul mate, Gottesman, wrote his Ph.D. thesis on the heritability of personality. These mid-sixties, genetics-oriented efforts were a decade before the publication of E. O. Wilson's landmark *Sociobiology* and fifteen years before Bouchard began his study of reared-apart twins. For that day, Scarr and Gottesman were two lonely genetics voices crying in the behaviorist dreamland.

Much of the prominence Scarr has earned over three decades in child developmental psychology derives not only from her research projects and papers but also from her voracious reading of the literature, her broad knowledge of all the most relevant studies, and her painstaking analyses of this sea of data. Her plunge into the cauldrons of race and I.Q. was not only motivated by her appetite for battle. "It is no accident," she wrote in 1981, when at Yale, "that I chose to study genetic differences in behavior—even the possibility of race differences. From my educational and social background, those seemed to be pressing social issues, ones that had been neglected in scientific research."

Politically she describes herself as "a mainstream academic liberal." Since her senior year at Vassar, she has been a member of the American Civil Liberties Union, and throughout her life she has supported feminist and other causes dedicated to a more equitable society.

In spite of her belief in the power of genes, she is highly skeptical, as both a scientist and a woman, about genetic generalizations, especially those concerning group differences, and is acutely aware of the injustice such generalizations can inflict on individuals. For a close-to-home refutation of such inferences, she cites the studies that say men are better at math than women. Then with the pride of an intellectually perverse mother, she points out that her son is poor at math, while her daughter is a whiz.

Although an active liberal, Scarr feels strongly that her political positions should not intrude on her science—nor that other people's politics should influence hers. In 1981 she wrote: "There is no more dangerous idea than the thought that someone, somewhere, can determine what I can study and say about my research." Her career as a scholar reveals an overriding interest in learning what *is*, what the givens of the human condition may be. Only after arriving at a valid understanding of reality, Scarr feels, can the scientist then apply this knowledge to implement a political vision for a more just world—or hope that others will utilize the knowledge the scientist has helped develop to beneficial social ends.

With this stance she puts herself in direct conflict with those whom she feels, in *their* visions for a more just world, have decided what the givens *should* be, then try to control and bend knowledge to that vision. In 1981 she discussed the social injustice that resulted from human variation but said, "It is the suffering that should be addressed, not the genetic differences denied."

Scarr has not shied from the areas of greatest controversy. In 1977 she and a colleague, Richard Weinberg, did a study of transracial adoption—black children adopted into white families. The study provoked fierce responses even though the results were mostly favorable to blacks. She found that the black children raised in white families logged I.Q. scores that were higher than the average for the white population—a mean I.Q. of 106. This was surprising, since in the U.S. black children consistently scored lower than white children—which was, in fact, the disagreeable statistic that had been the basis of Arthur Jensen's inflammatory paper and later *The Bell Curve*, both of which attributed this discrepancy to genetic differences between races. The Scarr-Weinberg study seemed to say that given decent upbringings, blacks are smarter on average than whites. Confusing things still further, even with the impressively enhanced performances, the adopted black children consistently scored lower than their white siblings, which would indicate that the white kids in the adoptive homes were well above the mean in I.Q.

The entirety of their data convinced Scarr and Weinberg that the explanation was not that the white, middle-class environment made the black kids smarter, but that the I.Q. tests that suggested such a result are biased toward the white culture. To be sure, the blacks tested better with a white upbringing, but this did not indicate I.Q. enhancement from that upbringing, as might be assumed, only that it offered better preparation for I.Q. test questions. She appeared to be slyly suggesting that intelligence was an entity separate and independent from I.Q. scores, a conclusion that surely delighted the small band of radical environmentalists who claimed there was no such thing as intelligence, that I.Q. tests were bogus instruments, measuring nothing—except the effects of a white, middle-class upbringing.

The angry reaction to the Scarr-Weinberg paper, in spite of its politically welcome conclusions, indicated that the mere act of addressing racial differences, even if none was found, was forbidden in some quarters. This was just the sort of intellectual taboo that was for Scarr a red flag. In 1981, twelve years after Arthur Jensen ripped open the genes-versus-environment debate with his paper on black-white I.Q. differences, Scarr edited a book of essays courageously titled *Race, Social Class and Individual Differences in I.Q.*

To encompass the debate, she invited the most prominent people on both sides to submit chapters—Jensen himself and such opponents as Leon Kamin, Noam Chomsky, Gerald Hirsch, and others of the stop-genes brigade. Scarr also included the results of her black-adoptees study and a number of her earlier papers, with a sampling of opposing commentaries. As the book's editor, Scarr maintained an impressive balance, only revealing her position when defending her own work. In her no-holds-barred rebuttal of Jensen's criticisms of her interracial adoption study, Scarr says, "Only the mother of such a study could love it enough to protect it from the arcane statistical threats he presents." Regardless of which side you were on or how weighty your credentials, it was clearly unwise to mess with Scarr.

Because Scarr's book on race was a compilation of papers written over many years, it revealed a substantive shift in her thoughts about the importance of the environment in development. In a chapter that first appeared in *Review of Child Development Research*, published in 1975 by the University of Chicago Press, she wrote: "The myth of heritability limiting malleability seems to die hard. Until recently, the importance of the genotype's reaction range was underestimated; it provides alternative phenotypes for the same individual, depending upon crucial environmental factors in the development of that individual. There is no one-to-one correspondence between genotypes and behavior phenotype, regardless of the heritability of a characteristic." The scholarly lingo translates to: A child can turn out many different ways, depending on its environment.

This strong environmentalist point of view seemed to have been reversed by the time Scarr wrote a 1978 paper, again with Weinberg, "Family Background and Intellectual Attainment," which appeared in the *American Sociological Review* and which she included in the latter part of her book on race and class. The authors wrote, "The conclusion that we feel is justified by our data is that intellectual differences among children at the end of the child-rearing period have little to do with environmental differences among families that range from solid working class to upper middle class."

Looked at more closely, these two views are not so much in opposition as they are bridged by a major qualification. The first is saying that the environment can alter genetically influenced traits in important ways. The second says that within a broad range of environments, the inheritance of intelligence (also the topic of the first paper) is not affected by the environment. At first glance it sounds as though Scarr is switching from saying the environment can have an important effect on genetic expression to saying the environment has no effect. But she is not saying the environment can have no effect, only that it rarely does—that within a broad range of environments, the genome will express itself unaltered. This was the same conclusion Bouchard was drawing from his reared-apart twins.

Despite Scarr's near dismissal of the environment except in extreme instances, she ended her book with a strongly anti-Jensen assertion: "So far, I see no evidence for the hypothesis that the average difference in intellectual performance between U.S. whites and blacks results primarily from genetic racial differences." It was an interesting agreement with Leon Kamin, who felt that unless the evidence was 100 percent foolproof, an impossibility with studies of humans, no conclusions could be drawn. If Scarr's position appeared to be antigenetic she counteracted this impression in the book's final paragraphs; she complained of the hypercritical, microscopic scrutiny given the behavioral genetics data and of the critics' failure to see a hardwood forest because of a few scrub pines.

In 1986 Scarr was elected president of the Behavioral Genetics Association, and five years later she was elected president of the Society for Research in Child Development. For her presidential address for the latter organization, she wrote a major statement on her overall views on human nature that, in their forthright downgrading of the family-rearing environment in a child's development, upset the most fundamental tenets of her field and, not surprisingly, produced furious reactions. Her talk was not just a behavioral genetics manifesto; it was a declaration of war on the obstructionist tactics of the environmental traditionalists.

The address's opening shot was her endorsement of the notion, first advanced by R. Q. Bell in 1968, that children play a large role in constructing their own environments and a refinement of that idea— that parents respond in specific ways to different children. This was an immediate gauntlet tossed at traditional psychologists since it deposed the environment as the monolithic, external force it had been considered to be for decades and replaced it with an environment that, to a large degree, was molded by innate differences in the children themselves, which is to say by their genetic endowments. Genes not only make the kids, she was saying; they have a big hand in creating their own environment as well. In a succinct summation of children's influence on their environment (or more succinctly, their selection of it), Scarr said, "Parents don't buy their kids heavy metal."

She then cited behavioral genetics studies that indicate variations in children that are not genetic but that are instead caused by *nonshared* environmental factors, that is, factors specific to the individual child but not to his or her siblings. Since these one-child factors explained most of the variation not explained by genes, environmental factors common to the entire family were relegated to insignificance. This brushed aside such longtime favorites with traditional psychologists as socioeconomic status, ethnicity, and educational levels. With her gloves now off, Scarr said: "Being reared in one family rather than another, within the range of families sampled, makes few differences in children's personality and intellectual development."

This assertion was guaranteed to infuriate the environmentalists, all the men and women who had devoted their lives to studying differences in family-rearing environments. Her sweeping, stable-cleaning assertions on the unimportance of family always stressed her crucial proviso: as long as the rearing environment fell within the "normal range" for that species. Without citing any studies, Scarr defined the normal range for human infants as "environments that include protective, parenting adults and a surrounding social group to which the child will be socialized." Not a word about Dr. Seuss books, scout camp, or Little League. Extremely broad, her definition of normal range seemed to embrace every rearing situation except those of children who are locked in closets, starved, abused, and dropped on their heads. It did not specifically address such dire environmental factors as poverty, malnutrition, understimulation, and drug-addicted parents, but the implication was that such conditions would be considered extreme, outside the normal range.

Broadly inclusive as was her definition of okay environments, it was another broadside at prevailing psychological orthodoxy, which held that variations in rearing techniques can have indelible effects on children. Indeed, philosophies of child-rearing had been founded on that very premise, important careers built upon it, and best-selling books written about it. None of these rearing variations was important, Scarr claimed, and hammered home her point by saying, "Most families provide sufficiently supportive environments that children's individual genetic differences develop."

After elaborating on this idea, Scarr then attacked the studies of environmental effects on children by saying they never factored out the degree to which an environment might have roots in the parents' genes, which of course the parents share to a degree with the child. She gave the example of a book-filled home of an upper-middle-class family, a situation that had been previously taken as an environmental element plain and simple, and pointed out that the books were probably reflections of the parents' genes and not strictly environmental. If a child from such a family grows up to love books, there is no way to sort out whether the love stemmed from the book-lined environment or from inheriting book-loving genes from the parents. With this addition to Scarr's system, we now had the environment as part function of the parents' genes, in addition to being part function of the children's genes. In the old nature-versus-nurture paradigm, nurture turned out to be full of nature.

Stating her "normal range" of environment concept in a different way, Scarr said that, "ordinary differences between families have little effect on children's development, unless the family is outside of a normal developmental range." Citing a study by D. C. Rowe at the University of Arizona, Scarr added, "Good enough, ordinary parents probably have the same effects on their children's development as culturally defined super parents." Perhaps fearing she had been too hard on conscientious parents, she added that this finding meant parents no longer had to knock themselves out for their children. Maybe even better news for low-energy parents was that Scarr's theory absolved them of guilt if things turned out badly. Instead of kicking themselves for screwing up, parents could say instead they had been dealt some bad seeds.

Also in her address, Scarr commended reared-apart twin studies and referred to the Minnesota findings as "startling"—citing in particular the high concordance rate of .76 in I.Q.s of Thomas Bouchard's reared-apart twins. She rubbed salt into the now raw sensitivities of her environmentally oriented audience by saying, "The remarkable studies of MZ twins reared in different families challenge many cherished beliefs in developmental psychology, but fit nicely in the genotype-environment theory," the theory that Scarr was advancing.

She also referred to four studies (one by herself and Richard Weinberg) that came to the same conclusion: In comparisons of biologically unrelated siblings raised in the same home (adopted brothers and sisters), no correlation at all could be found in their I.Q.s. In spite of such powerful evidence of the rearing environment's lack of effect on I.Q., a large number of the angry responses to Scarr's paper fell back on the shopworn criticism that reared-apart twins are often raised in "similar" homes, thus explaining their similarities in I.Q.s. Those using this argument failed to explain how vague similarities of two unconnected homes could have a major equalizing effect on I.Q. scores when the same home (for unrelated siblings) had none.

Throughout her career Scarr has stressed the importance of coming to understand the sources of individual differences and deplored the environmentalists' insistence that no innate differences of any importance existed between humans, that we were all *made* different by our environments. Because of the many years this article of faith ruled psychology, she feels the study of individual differences lags far behind where it should be. This is particularly lamentable, Scarr says, because in "the situations to which we want to generalize—presumably social problems like crime, alcoholism, learning disabilities—humans behave very differently and we must learn why."

IN TERMS OF INTELLECTUAL VIGOR and prodigious research output, Robert Plomin is one of the Young Turks of behavioral genetics. He even looks like one—bearded, six foot three, given to cowboy boots. On the other hand, in terms of his scientific caution and measured view of the genes-environment contribution to human behavior, he is, in his late forties, an elder statesman of the field. Throughout his twenty-year career, he has instigated or participated in a number of the most important twin and adoption studies—primarily those emanating from the University of Colorado, where he was on the faculty, and the Swedish Adoption/Twin Study. The latter was a collaboration between the Karolinska Institute in Stockholm and Penn State, where Plomin is currently a professor of psychology.

Paradoxically, Plomin feels obliged to the Catholic school he attended in Chicago's inner city for his genetic-evolutionary outlook. "When I was about ten," he told me during an interview, "I had to do a show-and-tell for school. I had been reading a book on evolution, one of the Time-Life series on biology. It talked about Darwin in the Galapagos and the slightly different variations of one species of bird on the different islands. I felt sure that God didn't create one bird for one island, a slightly varied bird for another. At school, I merely said, 'Isn't this interesting?' But I barely had the book out of my school bag when the nun whisked me out of class and down to the principal's office. It was still a mortal sin for a Catholic to believe in evolution. I got kicked out of school. It was a case of show, tell, and get bounced. They didn't say 'Here is why you shouldn't believe in evolution.' Their strategy then was just to pretend it didn't exist. It was an important event for me—also informative.* I thought, if these people are this way about an issue like this, I better be skeptical about everything else. Instead of being brought in line, I went very much the other way.

"At college, I decided to go into psychology and happened to attend one of the few grad schools in the late sixties that had a course in behavioral genetics—the University of Texas—and it was required. The department was under a very prominent psychologist, Gardner Lindzey, who was later at Harvard. At Texas, Lindzey hired two behavioral geneticists, which may not sound like a lot, but it was probably two more than most universities had then. Lindsay said he felt the future of psychology was in behavioral genetics. Almost no one of his eminence felt that way at that time. I was blown away by the course. I felt it was so powerful."

Since graduating, Plomin's career has soared, making him one of today's most respected behavioral geneticists. While still enthusiastic about the burgeoning genes-behavior discoveries, Plomin, unlike many behavioral geneticists, is also fascinated by the large degree of environmental influence that remains after genetic influence is factored out. Because the heritability of many behavioral traits is in the modest 30 percent to 50 percent range, Plomin, while acknowledging the importance of these findings, feels much still needs to be learned about the remaining 70 percent to 50 percent, that is, the environmental elements that affect the developing human.

While other genetic converts have rushed off in pursuit of heritabilities, Plomin has devoted much of his attention to the altered view of the environment that resulted from the new understanding of genetic influences. "Behavioral genetics research," Plomin said, "has revealed as much about environmental processes as it has about heredity."

A book he coauthored with his wife, Judy Dunn, also a psychologist, entitled *Why Children Are So Different*, was a diligent search for elements of both the environment and genes that caused such wide variation among siblings. Plomin and Dunn reasoned that because of the modest amount of genetic influence on most personality traits, the rearing environment shared by brothers and sisters should be major

*In 1996 Pope John Paul II softened this position by stating evolution was "more than just a theory."

determinants of their personalities, and these personalities should be far more similar than they generally turned out to be. This anomaly led Plomin and Dunn to the conclusion that the largest influences on personality are the aspects of the environment *not* shared by siblings.

Plomin and his wife cited the abundant evidence to back up their hunch that the shared environment, the overall rearing setting, had negligible effect on a child's development. This was from studies—primarily the Texas adoption project—that showed little correlation of traits among *unrelated* siblings—adopted brothers and sisters—raised in the same family. Random, unrelated genes growing up in identical rearing environments produced near-zero correlations in I.Q. This held true in most of the traits measured.

When the degree of genetic variation was calculated among such biological strangers reared in the same home, it was low enough to leave a large amount of variance unexplained. This left but a sole possibility, the one Plomin and Dunn arrived at: aspects of the environment particular to each individual in the family. They might be events or conditions experienced by only one child—a prolonged illness, for instance, or harsh treatment by a parent. It might also be a prenatal event—oxygen or nutrient deprivation, bacterial infections, all sorts of happenings that because they are nongenetic are lumped under the heading "environment." And because they happen to only one child they are called "nonshared environmental factors."

Another variation of the environmental theme struck Plomin and Dunn, as it had Sandra Scarr. Children affect their environments, often producing "customized" environments not shared with siblings. An example of this would be a naturally unruly child who would confront very different parents than his or her well-behaved siblings (angry and scolding as opposed to smiling and loving). Pushing this concept even further, a profoundly spiritual child might choose a monastery or convent, in which case he or she would have altered his environment completely.

Plomin and his wife were also leaders, along with Scarr, in developing the concept that environments are not separate and distinct entities in which genes express themselves, but rather are to varying degrees products of those genes. A child growing up in a musical home or a religious one might become musical or religious because of his upbringing or because he shared with his parents predisposing genes.

A child growing up in a port city might well attribute his or her love for the sea to his upbringing. But the new view raised the possibility that his parents may have chosen this location because of sea-loving genes that the child inherited. If the latter explanation were true, the seaside was not a genetically neutral environment but a genetically loaded one. No sooner had the field of psychology begun to recognize the importance of genes in the emerging human, it became apparent that the genetic action did not stop at the human but spilled over into the environment as well.

This genome-phenome-environmental interaction has been developed most fully by the prominent Oxford zoologist and genetic theorist Richard Dawkins in his book *The Extended Phenotype*. In it, Dawkins views birds' nests and rabbits' holes as expressions of those animals' genomes, all part of their phenotypes or, as he puts it, extensions of their phenotypes, just as birds' wings or rats' tails are. In his strongly antibehavioral genetics book *Biology as Ideology*, Harvard population geneticist Richard Lewontin expresses the same idea: "[Organisms] actually construct their environments out of bits and pieces. In this sense, the environment of organisms is coded in their DNA. . . ." Lewontin does not make clear how fish create sea water or what bits and pieces humans use to construct the air we need to survive, but he agrees with most observers of environment-organism interactions that the former is affected, sometimes to a large degree, by the latter. (Without earthworms, dirt would not be dirt, and so on.)

This broader view of the environment has the effect of chiseling out hunks of what had previously been considered environment, pure and simple, and shifting it back to the genetic side of the fence. Or worse, it would declare certain aspects of the environment a complex mix of genes and environment, which would seem to put the matter back where it all began. When the behavioral geneticists speak of 40 percent or 50 percent heritability (derived from MZ-DZ comparisons), they are speaking of that portion of a trait's variation, for a given group of people, that comes from their DNA; it does not necessarily speak about the contribution made to that trait by effects of the gene-influenced portion of the environment—that is to say, indirect gene influence. Nor does it speak about the effects on the environment by genes shared with other humans in the vicinity, parents for the most part, genetically related humans who may have had large influences on the rearing environment. These are influences that may have stemmed from the adults' genes, which the child, to a degree, shares. All of these influences are ways in which genes can *indirectly* act on a developing child, but so far such effects have not been measured. This omission has the effect of making the heritability figures lower than they actually are.

Yet another point about the environment renders the entire subject even more muddled. When psychologists use the term *environment*, they are referring to *any* influence that is not genetic. As mentioned above, that includes prenatal influences that can be devastatingly powerful. Such womb conditions as fetal position, nutrition, hormone infusions, and viruses are all "environmental" influences but can inflict permanent behavioral changes onto the developing fetus.

Dean Hamer, who made a major breakthrough in 1993 by locating a DNA anomaly in a group of homosexuals he tested, told of the range of effects that could occur in the prenatal environment: "The environment includes whether you were lying on your right side or your left side in the womb, and a whole parade of other things."

Now, with the rapid advance in understanding about these many potential prenatal influences, the term *environment* would seem to have been broadened into near meaninglessness. Traditionally among psychologists, the environment meant family settings, schools, neighborhoods, national culture. Now the term also includes a "parade" of womb events over which no one has much control. Since the environment's assumed malleability is so central to the critics' opposition to a biological view of behavior, the broadened understanding would seem to have stripped the term of much political resonance. With the hard-to-reach prenatal influences coupled with the realization that genes affect postnatal environments, the term "environment" is a far more biologically loaded term than realized up till now.

The tidy idea of nature on one side and nurture on the other is finished, just as is the idea that one or the other of them calls all the shots. Fortunately, Plomin, Scarr, Kagan, and others are undaunted by this new complexity. They feel that the clearer understanding now enables them to weigh an environment with greater accuracy than before, to break it down with greater precision than had been possible when its biological component was ignored. Their goal is to zero in on the environmental variables that most affect developing humans and to ascertain the degree of influence. This, in turn, will enable them to assess more accurately prospects for beneficial interventions.

Many psychologists, on acceding to the point about genetic effects on the environment, would throw up their hands and declare that the new perception rendered the entire genes-environment interaction too tangled for productive study. Robert Plomin, on the other hand, is the sort of scientist who is excited by new complexities and the new challenges they bring to his field.

Like other psychologists who have focused on the DNA underlying behavior, Plomin has had to educate himself in Mendelian genetics and molecular biology. With two of the most distinguished pioneers in the field, J. C. Defries and G. E. McClearn, Plomin has written a basic textbook, *Behavioral Genetics: A Primer*, which lays out for students the fundamentals of molecular genetics, as well as the advances in recent years made by the observational studies of twins and adoptees, the area they refer to as quantitative genetics. While still involved in studies of the latter, the more traditional domain of psychologists, Plomin has joined many of his colleagues in turning his attention to DNA itself.

He is convinced that rapid advances in molecular-biological screening techniques will revolutionize behavioral genetics, perhaps in the same way that behavioral genetics revolutionized psychology. "Just fifteen years ago," he wrote in 1990, "the idea of genetic influence on complex human behavior was anathema to many behavioral scientists. Now, however, the role of inheritance has become widely accepted, even for sensitive domains such as I.Q." Plomin worries that the pendulum may have swung too far toward biological determinism and that people will overlook the important role the environment still plays in many aspects of personality and behavior. He cites the 40 percent schizophrenia concordance in identical twins, a figure that leaves a hefty 60 percent of the cause from the environment.

He is greatly encouraged by the many prominent developmental psychologists like Jerome Kagan who are now using genetic strategies in their research, men and women who hold no theory-grounded expectations for either biology or the environment. They root for neither but believe the answer lies in a fascinating, little understood, interaction between both. Plomin's greatest hope, he says, is that "the next generation of developmental psychologists will wonder what the nature-nurture fuss was all about."

UNTIL THE MINNESOTA separated-twins study issued its papers, most of the findings of genetic links to behavior had been derived from fraternal-identical comparisons of raised-together twins. The assumption of these studies was that if identical twins have more personality similarities than fraternal twins, their rearing settings being equal, the explanation must lie in the genes. To date, there have been large number of these studies involving large numbers of twins, and they all came to the same conclusion: Behavior is influenced by genes. Because of this, the antigenes critics have expended much energy on discrediting the study model itself.

Their main criticism of this approach, voiced repeatedly to dismiss the genetic evidence, is that identical twins are brought up to *be* more alike so, in effect, *become* more alike. Greater degrees of similarity for identical twins measured by the tests do not, as claimed, indicate genetic influence but rather indicate the greater social pressure on them to be more similar. Since this was a plausible criticism, at least until Bouchard's reared-apart twins produced the same results, beleaguered behavioral geneticists went to great pains to test it empirically. With every test they have found the proposition to be untrue.

When Kenneth Kendler was at work on a massive study seeking a genetic component to schizophrenia based on identical-fraternal comparisons, he tried to preempt the usual debunking effort by publishing a paper in the *American Journal of Psychiatry* that summarized a number of earlier experiments that tested the hypothesis of MZs and DZs having different upbringings even though they were reared in the same home. All the studies found this phenomenon of different treatment given MZs than DZs, to the small extent it existed, had no significant effect. In spite of this battery of disproof, the argument continued to be raised, and is to this day.

The studies are interesting for two reasons. First, they demonstrate the enormous pains behavioral geneticists have repeatedly taken to answer their critics and prove the soundness of their research. Second, because of the studies' effectiveness in answering criticisms, and because of their ineffectiveness in putting to rest this particular objection to DZ-MZ comparisons, they reveal the critics' imperviousness to empirical proof, or even worse, their obliviousness to it. For this reason, I feel it is worth examining in some detail the rebuttal studies Kendler cited. They also happen to be quite ingeniously constructed.

Before presenting the data on different treatment of identical and fraternal twins, Kendler examined the possibility that there was an element of twinness itself, or perhaps just of identical twinness, that might lead to the trait he was interested in, schizophrenia. He cited several studies that compared the rate of schizophrenia in twins to that of the general population. None found any significant difference.

He moved quickly to the main complaint: more similar upbringings for identicals than for fraternals. The first study utilized the difficulty, in some cases, in determining zygosity—whether a twin is identical or fraternal. Because this is sometimes outwardly deceptive, a number of twins have, over the years, grown up misdiagnosed. By finding a group of such "false identicals" and comparing them with "real identicals," researchers could determine whether or not being treated as identicals had any effect on the fraternals. Similarly, by measuring identicals who were mistakenly raised as fraternals—that is, "false fraternals"—researchers were able to determine if the results were consistent with the type of twin they were *perceived* to be or with the type they actually were. Either way, the twins invariably tested as the kind of twin they actually were, not as the kind they were thought to be.

Said in another way, if identical twins through some misperception were brought up as fraternals, as adults they tested in I.Q., temperament, personality inventories, and other areas with the same high rate of concordance as identicals, not as fraternals. The opposite was an even more telling refutation of the point. Occasionally, fraternal twins looked as much alike as identicals and were brought up as such. In spite of this misapprehension that existed throughout their childhoods, and receiving the upbringings identical twins receive, as adults they tested as the fraternals they were, with far lower concordances than bona fide identicals.

Another set of tests cited by Kendler found a number of identical twins who had been brought up to be as similar as possible, whose parents took every opportunity to emphasize their sameness—matching outfits, haircuts, toys—then compared these twins to identicals whose parents made every effort to *minimize* their sameness. The testers found plenty of both. The opposite upbringing styles had no noticeable effect on the twins' similarities when measured later in life. Again, the possibility of distorting effects had been sought but not found.

The different upbringing proposition was tested in still another way. Kendler cited a body of research that examined the possibility that twins highly similar in appearance might elicit different treatment from their environment. This could have an effect on their development and therefore skew efforts to sort out genetic from environmental influence. Here again, test results of high look-alikes were compared with the results of low look-alikes, and there was no appreciable difference between the test scores. An appearance of sameness apparently had no effect on the overall degree of sameness.

All in all, Kendler cited nine studies that tested for possible bias in identical-fraternal comparisons. All found no validity to the supposed weaknesses in the study design. Also, to avoid any accusations of lax self-policing on the part of the behavioral geneticists—rather like the Justice Department investigating itself to stave off a special prosecutor—Kendler, like any good scientist, laid out the basic data of the earlier studies in great detail and cited the relevant papers.

In his summary of these heroic efforts on the part of the behavioral geneticists to meet this frequent objection of the environmentalists, Kendler made no mention of the complete substantiation these studies have received from the Minnesota and Swedish reared-apart twin studies, which lack the potential pitfall of different MZ-DZ upbringings in the same home. He laboriously showed that the one complaint has no basis in fact. It would seem to put to rest once and for all this one complaint and force the critics to find different ones.

This was not to be the case. For more than ten years after Kendler's paper, opponents continued to cite the possibility of different upbringings given identicals as opposed to fraternals as invalidating twin studies. As late as 1994, the objection was raised in the pages of *Scientific American*. Sometimes the criticism is not alluded to directly. When other critics referred darkly to the "seriously flawed" nature of twin studies that compared monozygotic with dizygotic twins, more often than not the unnamed flaw turned out to be the one Kendler and others had refuted a decade earlier. And there is no possibility the critics who keep resurrecting this charge are unaware of the refutation.

Each time the flaw is cited in print, a weary behavioral geneticist will write a letter to the editor pointing out the research that obviates the complaint, but the critics continue to make it year after year.

As an outsider, I came into this field believing scientists were simply truth seekers, men and women dedicated to discovering the functioning of the world around them, to understanding the givens. I saw them as driven by profound curiosity. It was, therefore, disheartening for me to learn that many scientists with broad reputations do not place truth at the top of their agendas and react in sadly unscientific ways when confronted with evidence they feel threatens their ideological positions. Aware of the scientific rules, they first attempt to discredit with counterarguments, but when these are shown empirically to be invalid, they simply pretend that the evidence they were unable to shoot down doesn't exist. Such selective memory permeates the behavioral genetics debate. In the nonscientific world we have a word for such behavior: *dishonesty*.

TEN

THE OTHER END—
SEARCHING THE DNA

IN A SUNNY GROUND-FLOOR LABORATORY in the Clinical Research Building at the University of Pennsylvania, Dani Reed, a young woman who looks a little like Pia Zadora and not much like a post-doctoral research fellow in molecular biology, spends her day extracting DNA from human blood cells, replicating it about a million fold through a process known as the polymerase chain reaction, then "running gels." This means she separates the nucleic bases that make up the segment of molecule, segregating them by an ingenious process known as electrophoresis. Using the electrical charge in all DNA—as if things weren't complicated enough—the bases are slowly drawn through a gelled substance (British geneticist Steve Jones has made do with strawberry Jell-O). Because the varying size of the bases causes them to travel at different speeds, the gene fragment is eventually separated and readable.

I was visiting the laboratory to learn as much as possible about the DNA scanning process, and Dani patiently explained the operation while she worked. At a conference at Cold Spring Harbor some months later, the sponsors hoped to educate the group and allowed us to run gels. The journalist with whom I was paired had done the operation before and grew impatient with the tedious process. His fidgetiness made me think of Dani and the hundreds of others of highly educated people around the globe quietly spending their working days repeating this same operation. For genetic research you clearly need knowledge and the brains to assimilate it, but you also need patience. Lurking within the gel plates with their amplified strands of DNA are nothing less than most of life's mysteries, perhaps all of them. To many knowledgeable people, these chemicals *en gelée* are the day's most important scientific frontier.

As I was being carried away by the epic significance of it all, Dani showed me a fingertip glob of translucent substance that she said was a few million molecules of the DNA she was examining.

"That's what DNA looks like?" I asked.

"That's it," she said happily. "Looks slimy, doesn't it?"

The DNA on which Dani and her colleagues were working came from a group of people who were the subjects of a study on obesity. The DNA was being scanned in the hope of finding an irregularity common to the obese, one not found in the DNA of nonobese people. Such a discovery might indicate a gene that causes, or at least contributes to, human obesity. The search required endless DNA scans. Such is the drudgery of gene mapping. Similar work—all of it in search of genes related to behavioral problems like addiction, depression, and violent aggression—is proceeding in at least a hundred laboratories around the world.

For all the monotony of the work, gene mapping is now the glamour end of the business. When Bouchard, Plomin, Kendler, and all the others devote enormous energy and money to their observational studies, they are basically looking for numbers, for heritability figures. Years of such effort have produced a series of two-digit figures that indicate the mean heritability of a trait. Although these numbers have been powerful enough to turn the field of psychology upside down, they are general mean figures for a population and lack the scientific precision of locating a subinfinitesimal bit of nucleic acid that might condemn an individual to a lifetime of depression, retardation, or obesity.

A number of considerations go into selecting a trait for a DNA search. Obesity is on the border between physical affliction and behavior, a border that is rapidly disappearing or being made irrelevant by behavioral genetics research such as this. Many might not consider being fat a matter of great medical urgency, certainly not as dire as being afflicted with cystic fibrosis or Huntington's disease. It is, however, an excellent trait for genetic research in that the condition is easily recognized and diagnosed, unlike traits like depression and alcoholism, which occur in many forms. Obese people, often hating their condition and eager to find a cure, are easily recruited. The ultimate goal is not a cure for obesity, although that would be greatly welcomed, but rather unraveling the mysteries of gene function. The N.I.H.'s Dean Hamer, referring to his 1993 study linking genes and homosexuality, said, "It is likely that finding a 'gay gene' will be remembered as a breakthrough not so much for what it explained about sexuality as for opening another door to understanding genetic links to many equally complicated human behaviors and conditions." Hamer's point is true of all quests for behavioral genes.

The director of the University of Pennsylvania obesity study is molecular geneticist Arlen Price, a slender, bearded man in his forties with a soft voice and an academic air. He was the Minnesota Study kibitzer who goaded his graduate school friend Nancy Segal to speed up publishing the twin data. Price has spent ten years researching obesity, four of them mapping DNA, and is optimistic that a recent discovery has brought his goal nearer. "It is not a major breakthrough," he told me when I visited his office, "but it's the kind of development that keeps us coming to work every day."

In November 1994, Price and his group feared they were scooped when a group at New York's Rockefeller Institute, also engaged in obesity research, announced that they had found a gene in mice that was directly tied to obesity. The gene underlies a signaling mechanism that tells the mouse's brain when enough food has been eaten. After a mouse consumes enough food, the gene, when working properly, initiates a biochemical chain of events that culminates in the brain's telling the mouse he is no longer hungry, to stop eating. In certain mice this gene does not function, and the result is obesity. The Rockefeller group had reason to believe that in humans the same gene might malfunction in the same way. (It was later found that the human version of the gene did not commonly have mutations similar to those in mice, dashing hopes for an obesity cure and leaving the field open to Price and his team. The animal-human parlay sometimes pays off, sometimes not.)

In their work with human DNA, Price and his team had already been zeroing in on the same genetic region as the Rockefeller people. In a model of scientific cooperation, the two groups immediately began trading information, vastly speeding up the progress of both. By December 1994, the University of Pennsylvania team was ready to announce that it had reached an intermediate finish line: They had linked the gene but could not demonstrate its role in creating human obesity.

Price is an example of the new breed of crossover behavioral scientist. Since his student days, he has shuttled between psychology and molecular biology, before those disciplines drew as close as they are now. He began his academic career in the sixties as a biochemist, then switched to psychology and obtained his Ph.D. at the University of Colorado. When the Jensen controversy erupted in 1969, Price was troubled by science's failure to answer definitively such fundamental questions as whether or not I.Q. was genetically based. He wondered if there might be a way to establish, with this or any trait, a genetic component that would be beyond argument, a DNA proof of some sort. He saw the unfairness of the attacks on twin and adoption studies and watched in dismay as these painstaking efforts were shot down by armchair critics, environmentalists like Leon Kamin who comb the data in search of potential flaws with which to discredit the findings or at least spread debilitating doubt.

Seeing how easy it was for the opponents of behavioral genetics to wound the most cautious studies, not by counterresearch, but by ad hoc sniping, Price despaired of the observational studies' ever satisfying the critics' objections, yet he remained convinced of a genes-behavior link. He decided he did not want to devote his life to devising ever more ingenious ways to plug possible holes, to knock himself out anticipating every conceivable vulnerability. He knew that even if such airtight studies could be created, the opponents would still claim to have found holes. After working in the field for a time, he sadly acknowledged that with behavioral genetics, as much energy had to be spent defending data as producing it. Convinced the observational studies would never be free from attacks, he made the switch to molecular biology, first by doing postdoctoral research at Stanford with eminent population geneticist Luca Cavalli-Sforza, by working on DNA-behavioral research with Kenneth Kidd at Yale, and then by starting a molecular laboratory at the University of Pennsylvania.

"I saw the whole thing continually collapsing into unresolvable arguments, so I grew envious of the molecular biologists down the hall, the people who worked with concrete experiments cloning genes, not as hypothetical elements, but as sequences of bases. I liked that their work was *not* controversial."

The wistfulness Price felt about hard science is not uncommon among those toiling in the soft terrain of psychology and sociology. There is even a term for the condition, "physics envy," which describes the feelings of all those who wish their observations could be codified into the numbers and formulae physicists use to record their discoveries. In fact, psychologists, in particular behavioral geneticists, have taken to their own brand of formula—rather like fortifying their conclusions with the chain mail of hard science. These highly complex constructions of letters and numbers that occasionally erupt in psychological papers can be daunting even to those with long experience in the field. But not everyone takes them seriously. An eminent psychologist, an elderly German, was asked how he dealt with these intricate environment-genetics formulae when he encountered them in a text. His face lit up and he said, "I hum them."

But instead of aping the hard-science vernacular in presenting their observation data, more and more behavioral geneticists were, like Price, jumping onto the hard-science bandwagon made available by the advances in molecular genetics. They are examining the DNA itself in an effort to isolate anomalies that can be positively identified as at least a contributing cause of specific human traits and idiosyncrasies. Neurobiologist Simon LeVay's image for the two approaches to behavioral genes has it that traditional approaches with twins and adoptees are looking at behavior "from the top down," while molecular biologists are tracking it from "the bottom up."

To many in the field, pinpointing the specific DNA for a behavior is the Holy Grail of behavioral genetics and a reason why more and more psychologists are lured to the molecular end of the quest. Finding a specific gene for a specific behavior would mean that after twenty years of circumstantial evidence from the observational studies—not to mention the two thousand years of casual observation of family traits and animal-breeding techniques—hard proof of a genes-behavior link would finally be on the table.

Finding and implicating specific genes would not, however, signal the ultimate untangling of the human-behavior enigma. As Thomas Bouchard said, "We know that genes make proteins, but that is many biochemical steps away from behavior. What lies in between is pretty much a mystery." Since Bouchard made the remark in 1992, significant advances have been made in understanding what lies "in between," but there is still much that is not yet understood. A lot of research under way is aimed at understanding the biological mechanisms that translate chemicals into behavior. With the combined efforts of biochemists, molecular biologists, and behavioral geneticists, advances are rapid. Together they are prying open the secrets of what they reverentially refer to as "the black box."

Although much still has to be learned, there is no doubt that pinpointing specific genes for specific behaviors is an important next step in unraveling the gene-behavior mystery. In addition to representing a giant leap forward in scientific understanding, the incontrovertible locating of specific behavioral genes—whose functions have been established, the experiments replicated, and the DNA sequenced— would force the consideration of genetic elements in all behaviors. With luck, it would go far toward silencing the obstructive heckling from the environmentalists. Because this is a totally different method of exploring the gene-behavior connection, the genephobic environmentalists will have to come up with an entirely new battery of "flaws."

BEFORE DELVING INTO the methods for locating behavioral genes, it is helpful to have some understanding of human DNA, the decade's scientific superstar. Books on genetics aimed at the general reader invariably have a brief course on DNA, including such fundamentals as the arrangement of genes along a chromosome, the way in which cells divide, and the way DNA replicates itself. When I began seriously investigating the field, I became somewhat of a connoisseur of these primers and came to appreciate the authors' valiant efforts to make the process appear, if not simple, at least straightforward and accessible. The sad truth is that it is of a complexity to daunt the most intrepid intellect and at the same time strike a reverential awe for nature into the most blasé and thrill-weary.

To have a comprehensive understanding of the present state of the science, one would need a solid grasp of molecular biology and biochemistry. A good grounding in physics would come in handy as well. As the miracle of it all unfolds, it might be a good idea to have some theology at the ready, if for no other reason than to stave off the inevitable question of how any process so magnificently intricate could come into existence. Actually, for this you only need a solid understanding of Darwin's principles, followed up by a reading of Stephen Jay Gould and Richard Dawkins, two reigning evolutionary theorists, who disagree on some points, but who are both masterful at clarifying the knotty ramifications of Darwin's simple but profound insight. Dawkins is particularly good at making plausible such complex evolutionary products as a bat's night-flying ability, and he does so without postulating a deity who fancied having a creature around who could fly in the dark without bumping into things.

Among the ramifications they illuminate is the simple fact of human variation and how essential tiny changes are to the evolution of different species. To assure this variation, of all the billions of sperm and egg cells that become humans, as well as the zillions that are produced but never develop, no two are the same. Identical twins, who are genetically the same, are not exceptions to this fact, as they come from the *same* egg and sperm cell. As chromosomes divide in the reproductive process (meiosis) and half the mother's DNA binds with half the father's—the genes shuffle their millions of elements to produce endless variations on the human theme. This process assures the species infinite opportunities for change, yet always sticking within genetic boundaries that guarantee a degree of species stability.

This roiling sea of variation is occasionally given a jolt with an outright mutation, a tidal-wave variation like six toes or smaller ears—anomalies that will probably go nowhere but into the evolutionary ash can. Of all mutations that occur, many of them, estimates run as high as 50 percent, abort before birth. If, on the other hand, a mutation appears in a living organism and gives it a procreational advantage, the change, through natural selection, will all but inevitably spread slowly throughout the species. If it is a highly advantageous mutation, in a brief million years or so it will probably become a permanent part of the species' basic equipment and lead to the extinction of all members of the species lacking it.

According to the fossil evidence, it generally takes vast periods of time for mutations to achieve specieswide success, but a 1994 book,

The Beak of the Finch by Jonathan Weiner, provides strong evidence that the evolutionary process can occur much faster—with the Galapagos finches, in a scant fifty years, even less. This has interesting ramifications for humans, in particular for the pessimistic idea that we evolved to our present form in an environment totally different from the one in which we now find ourselves. We are stuck, the thinking goes, with outdated evolutionary equipment and cannot hope to evolve fast enough to catch up. Weiner's book offers hope. I can't predict which human traits will be detrimental and which advantageous, but just on a wild hunch, I would bet on computer whizzes for winning the genetic sweepstakes. And I wouldn't rule out seven-foot basketball stars.

Another feature of many DNA primers is a reluctance to announce parts of the picture not yet known. I found all the books quick to reveal that each of the 100 trillion cells in the human body has a complete complement of that body's DNA. Each cell's DNA is broken down into twenty-three pairs of chromosomes. The DNA from just one set of these chromosomes contains all the information needed to make and operate a human and, if laid out in one continuous strand, would be six feet long. That's just the DNA from one of our body's trillions of cells. (British geneticist Steve Jones makes this incredible fact even more indigestible by pointing out that if the DNA from one human were stretched out in a line, it would go to the moon and back eight thousand times.)

Except for identical twins, the DNA in each cell is identical and particular to one individual, to the dismay of many criminals. Each cell's DNA, all twenty-three chromosomes, is made up of some three billion nucleic bases. Most of this appears to be nonfunctioning, "junk" DNA as it is called, but about 3 percent are operating genes. The total number of working genes is thought to be somewhere between 50,000 and 150,000. The wide spread in the estimate reveals a glimpse of the lingering uncertainties among scientists about fundamental divisions. On the other hand, considering that the DNA molecule's structure was only discovered forty years ago, it is amazing how much of the overall arrangement *is* understood.

As I waded through the textbooks, I was struck by the absence of any talk about the relationship of DNA molecules to chromosomes. The books all say that the double helix is a long, to-the-moon molecule and it is, at times in its life, wound yarnlike in the sausage-shaped spindle of the chromosome. No mention is made, however, of the number of these molecules in a chromosome. If the number of chromosomes in a cell is important enough for constant reiteration, why not the number of DNA molecules in a chromosome?

Sure I had nailed them for leaving out this elementary point, I popped my question to a prominent molecular biologist at the N.I.H. Slowly, he replied, "We think it is one molecule per chromosome, but we don't know."

Don't know? How are they going to find the one-billionth bit that causes depression or optimism if they don't even know the number of molecules in a chromosome? Adding to my confusion, when the scientists talk about cutting snippets of the DNA molecule, they also call the resulting fragment "a molecule." Like worms, a DNA molecule cut in half becomes two DNA molecules. Such sliding terminology does not inspire confidence in the nonscientist who prefers words to stay within definitions, especially one who is struggling to learn the rudiments of molecular biology.

A gene is defined as a sequence of nucleic bases that is self-replicating and from which a protein can be synthesized. Proteins are what make everything happen, what is more, they *are* everything. They have many forms and functions—messenger proteins, instructive proteins, timing proteins, and building-block proteins. The gene creates them and tells them their mission. They then spread the word to all relevant departments.

Richard Dawkins, in his book *The Selfish Gene*, offers the controversial view that the gene is the end product, the point of it all. Organisms like humans are mere vehicles for serving, generation after generation, the gene's mindless drive toward immortality. According to Dawkins, the relentless, exploitive gene uses our bodies to replicate itself and reproduce for future use as many similarly hospitable bodies as possible. The gene aims only to propagate itself and assure "its" survival in ever more numerous generations of descendants. As with most theories that reduce our cosmic status and render us less adorable (us? selfish?), this theory is hotly contested. Presumably for scientific reasons and not wounded pride, Stephen Jay Gould feels such a view of genes mistakes the messenger of change with its cause. The argument, however, seems more about the precision of Dawkins's metaphor rather than the genetic principle it seeks to describe.

The Dawkins picture of gene single-mindedness gets murky when you consider some 100,000 genes, all with different functions, using for their selfish ambitions the same "vehicle." And these vehicles that are merely serving the genes' ambitions would include every human—from Mother Teresa to Saddam Hussein—plus every other earthly creature, from microspira protozoa to sperm whales. Anarchical as this sounds, there is more than a little evidence that it is true, at least the part about genes having different purposes from one another and certainly from us vehicles. There they all are, pushing us this way and that, each doggedly determined to carry out its mission of self-perpetuation, sometimes in concert with their neighbors, sometimes in opposition—but all oblivious to everything but their own survival.

In the past year or two, the lid has been lifted a bit from the black box; and the peek has revealed a lot about the mechanics of genes and the way they function. When a gene goes into action, which happens in different cells at different times, it transcribes its information onto a molecule that it assembles from the free-floating sugars and phosphates in the surrounding nucleus. This is messenger RNA that carries the genetic news into the surrounding cytoplasm, where it hooks up to a molecule of transfer RNA, to which it delivers its information. The transfer RNA assembles itself according to this information and, depending on its form, proceeds to attract one of twenty amino acids to form polypeptide chains that constitute the proteins and enzymes that make up humans. (The word *protein* is chemically descriptive, whereas enzyme refers to function. Enzymes are proteins that act as organic catalysts.) When activated, RNA carries out the DNA's instructions to form a particular protein very quickly, at a staggering rate of 100 molecules a second. The same complicated process, using the same genetic code, occurs in every living organism.

Some of the genes in each human cell carry instructions for building arms, legs, and livers. The genes for creating the complex hemoglobin molecules are well understood, as are many of those creating various enzymes. But as an indication of how complex it all is, and as an indication of why it takes three billion base pairs to make an operative human, there are sixty different proteins in a human tear, two hundred in a drop of blood. To locate the specific genes for creating arms, legs, and tears is a complicated business. It is undoubtedly far more complicated to pinpoint genes that influence behavior.

In SETTING OUT on the grueling task of locating behavioral genes, the current procedure is to find families that have a high incidence of a recognizable and unusual trait or disorder, enough instances within one family to suggest genetic involvement. Then from as many family members as possible, the DNA is examined in the hope of finding a variation that is common to the afflicted members but absent in the nonafflicted. Because any one of the three billion DNA base pairs could be involved, the search is daunting.

In describing a hunt for a behavioral gene, a frequent analogy is searching for a leaky faucet known only to be somewhere on Earth. If a gene search can be narrowed to a specific chromosome, that is comparable to learning the leak is somewhere in Texas. There are techniques for narrowing the search somewhat—to tell you, for instance, that the problem faucet is in Dallas; but here the shortcuts end. Having arrived at the correct city, you must walk up and down its streets knocking on doors until the leak is found. Given the length and the thinness of the DNA molecule, a better analogy might be seeking a defective spike on the trans-Siberian railroad. In a sense, that is what Arlen Price and Dani Reed and hundreds of others do each day, walking the track, or the streets of Dallas, looking for an infinitesimal snippet of DNA that causes a behavior.

At the present time the most successful means of narrowing the search is the linkage study. This technique rests on the principle that if traits lie close together on a chromosome, the chances are extremely low they will separate during the DNA shuffling that occurs when offspring are conceived. The search is aided by the existence of markers, locations on the DNA chain that appear at the same spot on every chromosome and are sufficiently similar in all humans to serve as guideposts and sufficiently different to be identifiable as coming from one family. If the guide post and the trait appear together consistently in subsequent generations—freckles and diabetes, as a hypothetical example—the suspicion runs high that the marker and the target gene are too close to separate during the gene shuffling at inception. The two genes are said to be linked. The principle is that if two knots are a quarter inch apart on a ball of twine and the twine is cut every twenty-five feet, the chances are almost nonexistent that a cut will separate the knots. It's not impossible, just extremely unlikely. If the two traits repeatedly turn up together in generation after generation, the searched-for trait is said to be linked to that particular marker. Since the location of the marker is known, the rough location of the searched-for gene is now known as well.

But when linkage is discovered, that brings the gene-tracker only to Dallas. There are still all those streets and buildings to search. Still, closing in even to the degree of finding the right chromosome, or better, the gene's general location on the chromosome, is considered an important advance, as when the Huntington's chorea gene was linked in 1983 by Nancy Wexler and James Gusella. But as discouraging evidence of the gulf between linking and finding the actual gene, it took six research teams working at ten different institutions another *ten years* before they found the Huntington's gene itself in 1993.

Even having at last achieved the miracle of pinpointing the actual gene—Huntington's turned out to be, as suspected, a single, dominant gene—Wexler and her colleagues have no idea how it causes the disease, nor have they devised a strategy for disabling it. Still, few would argue that finding the specific gene put them no closer to a remedy.

OF THE VARIOUS LINKAGE STUDIES in recent years aimed at pinpointing genes for behavioral traits, the biggest breakthrough appeared to come in 1987, when a group headed by Janice Egeland from the University of Miami School of Medicine studied thirty-two Older Order Amish families in which a high incidence of manic depression was found. The Pennsylvania Amish were an ideal group for genetic study in that they were descended from the same forty families, they kept detailed health records, and there was little marriage with outsiders. The group had come to the melting pot in the nineteenth century and had not melted. They were also free of the alcohol and drug-abuse problems that can wreak havoc on diagnoses of amorphous conditions like depression and anxiety disorders.

When the Egeland group announced linking the depression-causing gene to a marker on chromosome 11, it was heralded on the front pages of the *New York Times* as a landmark genetic development. It was the first claim of hard evidence that an intricate behavioral disorder was inherited in a straightforward Mendelian pattern. The finding caused considerable excitement and looked as though it would bring back to life psychiatric genetics, a field all but comatose since the advent of Freudianism. The Amish study, however, enjoyed a far briefer period of triumph than did Freud. When another team from the National Institutes of Health tried to replicate the study, at the same time increasing the number of family members examined, they could not substantiate the earlier findings.

The same year, similar hopes were raised for locating a genetic basis for manic depression by a group at Columbia University headed by Miron Baron. Three Israeli families were found with an unusually high incidence of clinical depression. After scanning their DNA, the Columbia team announced they had located the genes responsible. This group, too, later had to retract their claim because of insufficient evidence. With the Amish study, the biggest blow came when two of the individuals considered well—no signs of the disease and lacking the incriminating stretch of DNA—later came down with depression. This seriously reduced the statistical weight of the earlier finding. The Barron study apparently had different methodological problems. With or without design flaws, both studies provided discouraging evidence of the complexity of the gene-behavior relationship and the elusiveness of pinning down direct biochemical causes. They also proved yet again a difference between studying humans and white mice.

In 1990 excitement again ran high when Ernest Noble, a psychiatrist at the University of California, and Kenneth Blum, a pharmacologist at the University of Texas, announced they had found a gene on chromosome 11 for alcoholism, or more precisely, "a reward gene" that heightens susceptibility to alcoholism and many other compulsive behaviors. Of a large group of alcoholics tested, an impressive 69 percent had a variant in this gene that was not present in 80 percent of the control group. When the discovery was made public, again on the front pages, the finding was taken as the long-awaited hard proof of a gene-behavior connection.

As with the Amish study, the victory unraveled quickly when other institutions could not replicate the findings. Convinced they were right, Noble and Blum argued against their detractors' methods—alcoholics in the control groups, for instance, or eliminating for medical reasons the alcoholics most likely to carry the gene variant—and they may yet be vindicated; but for the moment the study has been judged another behavioral genetics failure.

PREDICTABLY, THE OPPONENTS of behavioral genetics were de-lighted by these misfires, touting them as proof of zero connection between genes and behavior. Such a conclusion was unjustified. With the Amish study, the most casual glance at the high incidence of depression in certain families and not others, and among skipped gen-erations unknown to each other (a depressed child's grandfather was also depressed, for example), should dispel anyone's doubt that genes and heredity were somehow involved in the phenomenon. The Amish study's collapse meant only that searching along the three billion base pairs in the human genome, these scientists were unable to pinpoint exactly which chromosomal area or areas were responsible for the extraordinarily high number of afflicted people in one bloodline over a number of generations. But in the shortened press versions of the set-back, the impression was left that a genetic connection to depression had yet to be established. This was untrue; science had merely failed to locate the specific gene.

Many were beginning to feel the misfire problem lay in leaning too heavily on the assumption that only one gene was causing the problem under investigation. Understandably, behavioral researchers were envious of the genetic triumphs relating to physical afflictions like Huntington's and cystic fibrosis. Like Mendel with his peas, the scientists involved had been lucky to have targeted disorders for which only one gene turned out to be responsible. There was no guarantee such genetic simplicity would be true of all conditions; with behavior, the likelihood was high it would not. The elusiveness of behavioral genes was bringing more and more scientists to the view that behav-ioral traits invariably have more than one genetic cause—either differ-ent genes working in concert to bring on depression, for example, or totally unrelated genes causing the same disorder in different people. Your depression might have one genetic cause, mine another. A third person's might not be related to genes at all.

A clue to the search's difficulty is suggested by a statement of Richard Dawkins, who was speaking about the genes that construct the body: "The manufacture of a body is a cooperative venture of such intricacy that it is almost impossible to disentangle the contribution of one gene from that of another. A given gene will have many different effects on quite different parts of the body. A given part of the body will be influenced by many genes and the effect of any one gene depends on the interaction with many others." The complexity he describes may well be greater with behavioral genes. The concept of one-gene-one-behavior is the reductionism the critics quite rightly deplore.

Robert Plomin is among many who hold to the multigene view of behavioral traits and is quite sure this complexity explains the lack of success in implicating specific genes for specific behaviors. In an April 1994 article in *Science*, Plomin argued that all the evidence suggested that behavioral traits were not influenced by single major genes but by an array of genes, each with small effects. He views the single-gene approach as doomed to failure. While stressing the complexity, Plomin sees hope for progress in a different direction. "I'm interested in merging molecular genetics and quantitative genetics," he says. "That's what many of us are trying to do, not saying we think there's a single gene and we hope to stumble on it. But rather let's bring the light of molecular genetics into this dark alley and look for genes here. And that means we need approaches that will allow us to find genes that account for very small effects—not 20 percent of a trait's cause, not 10 percent, but less than 1 percent. There are ways to do that. Association approaches. The Human Genome Project will speed up this sort of research."

Plomin is aware this is a controversial area. "Some would say there's nothing to this. But they don't like it because it's so against the idea that if something is genetic, there must be a simple gene there. You can find it if you just measure correctly. They say the reason we're not finding single genes is that the psychiatrists and psychologists are not measuring them right. I just don't buy that. You can measure neurotransmitter levels and I don't think they are single-gene phenomena.

They are incredibly complex physiologically, and I'm sure that a lot of genetic variance contributes to it."

MANY BEHAVIORAL GENETICISTS have long been aware of the complexity of not only the gene-environment interaction but also the *gene-gene* interaction—the entire human motherboard making New York Telephone's mainframe seem simple and straightforward. Even before awareness of complicating factors, behavioral geneticists have known that human DNA is a sprawling and quirky vastness. But they take heart by reminding themselves that forty years ago we didn't know what DNA was. It is just beginning to be explored and understood. Few in the field see the failures as evidence they are not on the right track. In contrast, the critics' precipitous claims of vindication over the few failed efforts are seen by many geneticists as nothing more than strong evidence of the opposition's dread of hard scientific proof of the genes-behavior dynamic.

For the faithful, however, even the failures suggest the distance behavioral research has traveled in the past ten years. The picture lingers of scientists drawing blood from the fingertips of Amish and Israeli patients in search of the cause of their emotional problems and mental illnesses. The tableau stands in vivid contrast to another that served as the twentieth-century paradigm for mental therapy: a patient stretched out on a couch with an analyst seated to the rear taking notes. Together, the opposing vignettes provide eloquent testimony to the revolution behavioral genetics has brought about, not only in psychiatry but also in our entire concept of human behavior, sick or well.

ELEVEN

MOVING RIGHT ALONG
THE DOUBLE HELIX

IN 1991 EXCITEMENT again ran high that a physiological link to a behavior had at last been found, one strongly suggestive of a genetic basis to the trait—this time homosexuality. *Science* magazine published a paper by Simon LeVay (then conducting neurological research at the Salk Institute in La Jolla, California) that reported the discovery of a structural difference within the brains of a group of gay men.

LeVay had been intrigued by Dutch research that had found an anatomical difference between gay and straight males; a cluster of cells in the brain known as the superchiasmatic nucleus was nearly twice as large in homosexual men as it was in heterosexual men. LeVay, however, knew that this region of the brain had not been associated with sexuality and the variation might be an *effect* of homosexuality rather than a cause. He undertook to investigate an area that *had* been linked to sexual behavior, the hypothalamus—in particular, four regions of the hypothalamus. These regions, because they were known to be larger in men than in women, appeared to be sex-connected, so were seen as likely candidates for an anomaly linked to sexual orientation.

LeVay examined the brains of nineteen deceased homosexual men, sixteen heterosexual men, and six heterosexual women. He found that one of the four hypothalamic regions, INAH-3, was twice as large in the heterosexual men as in the homosexual men. Remarkable as this finding was, the study came under strong criticism, as all of the homosexual men and six of the heterosexual men had died of AIDS. To most observers' satisfaction, LeVay disposed of this criticism by pointing out that no significant difference could be found in the size of INAH-3 in the heterosexual men who had died of AIDS from the ten who had died of other causes. It seemed clear to LeVay that AIDS had no effect on the size of this portion of the hypothalamus, while homosexuality appeared to have a big effect.

Stunning as was LeVay's discovery, it was still vulnerable to the same fallacy that left the Dutch discovery inconclusive; that anomalous brain size might be a *result* of homosexuality rather than a cause. Until the exact origins of this brain irregularity could be determined, its significance was uncertain. There were two additional confounding possibilities: LeVay's sample was small and his finding had not been replicated. Aware of his discovery's tentativeness, LeVay was highly cautious in his conclusions, saying only that the anatomical difference he found between gay and straight men "illustrates that sexual orientation in humans is amenable to study at the biological level." It certainly shifted the entire subject of sexual orientation still further from the psychoanalyst's couch, ink-blot tests, and smothering mothers. Many hoped LeVay's surprising finding might also move homosexuality away from the guilt and shame that had long accompanied it.

With the concept in the air of an innate basis of homosexuality, strong corroboration arrived two years later. In July 1993, the *New York Times*, now understandably skittish about claims of specific genes-behavior connections, again heralded on its front page a major finding linking homosexuality to DNA. A research team at the National Institutes of Health led by molecular biologist Dean Hamer published in *Science* the results of a study in which they had established a genetic anomaly on the X chromosome of thirty-three out of forty pairs of brothers, both of whom were homosexual.

For years behavioral geneticists had been closing in on homosexuality. Earlier twin and adoption studies, including Minnesota's, had turned up persuasive evidence of heritability of this trait, and the issue appeared to be resolved in 1991 with the publication in the *Archives of General Psychiatry* of a twin study by Michael Bailey, of Northwestern University's Psychology Department, and Richard

Pillard, a psychiatrist at Boston University's School of Medicine. Recruiting 110 male twin pairs (56 MZs and 54 DZs) in which at least one was homosexual, they found that 52 percent of the identical twins were also gay as opposed to 22 percent of the fraternal twins. When these men had adopted brothers as well, only 11 percent of the nonbiological brothers were gay. While this suggested that the environment still played a large role in the development of homosexuality, the far higher concordance between the identicals than the fraternals was powerful evidence that genes were involved.

As with most observational studies of complex behaviors, the Bailey-Pillard study was vulnerable to numerous criticisms on recruitment, assessing gray areas of dubious sexuality and so on. The study was also damaged by an odd statistic: a 9.2 percent rate of homosexuality in the subjects' nontwin brothers. Inexplicably, this was a lower figure than that for adopted brothers who had no genetic connections; to fit tidily into the model it should have been higher, and roughly the same as the percentage for fraternal twins. Still the 52 percent for MZs as opposed to 22 percent for DZs was taken by most people as telling evidence of a genetic influence on homosexuality. Then with LeVay's riveting finding in the hypothalamus, homosexuality was ripe for hard DNA evidence, and the Hamer study appeared to have delivered it.

While designing his experiment, Hamer worked on the belief that if a genetic component to male homosexuality were to be found, it would be on the X chromosome because of a pattern of transmission he had observed through the maternal side. Analyzing the family histories of 114 male homosexuals, Hamer was surprised to discover a much higher than expected number of homosexuals among the subjects' uncles and male cousins, but always on the mothers' side of the family, never the fathers' side. This observation greatly strengthened Hamer's hunch that if the condition were genetically caused, the gene would be on the X chromosome that comes from the mother.

I was struck by the scientific community's years of obliviousness to this strong evidence of maternal transmission, which had long been apparent to people with homosexuality in their families. To me, it spoke volumes about environmentalism's stranglehold on thinking about behavior. If a gay man's maternal uncle and grandfather were gay, it would be written off as conditioning by some hidden rearing factors—a family passion for Judy Garland, perhaps, or season tickets to the Metropolitan Opera—if not simply a family tradition. (As in the old joke about the French duke: Fed up with an Englishman's boast about his ancient family, the duke said, "My dear fellow, when your ancestors were living in caves and painting themselves blue, my family was *already* homosexual.") For many years, such readily at-hand information as gayness on the maternal side, no matter how evocative of a genetic pattern, if inconsistent with environmental theories about a "psychological" condition like homosexuality, was rendered, if not invisible, at best anecdotal, meaningless, coincidental. To suggest a pattern of genetic *inheritance* was to brand oneself as scientifically ignorant.

Hamer and his research team, proceeding on his X chromosome hunch, examined the DNA of their sample and quickly found that the majority of the gay brothers, 83 percent, shared a region at the tip of the X chromosome that was unusual in an identical way. If the distinctive tip had occurred by chance alone—that is, if it were unrelated to homosexuality—it would have been shared by roughly 50 percent of the brothers since it must have come from one parent or the other. Surprisingly, the Hamer group made no effort to determine whether or not nonhomosexual brothers or other relatives of the gay pairs also possessed the same chromosomal tip. Still, the 83 percent concordance between the gay brothers was seen as proof of the DNA region's involvement in their homosexuality. It would be *related* to the homosexuality, Hamer stressed, but not necessarily the cause. He also stressed that his findings might relate to only one type of homosexuality; other types might exist that had altogether different causes, either genetic or environmental.

In the resulting paper, the Hamer group pointed out that the region on the long arm of the X chromosome where they had located the shared variation, while representing less than .2 percent of the human genome, was still some four million base pairs in length. This was a lot of base pairs to rummage through in search of the actual gene or genes involved. Hamer felt that locating them would require either a study with many more pairs of gay brothers or the complete DNA sequence of the region. Although he knew either strategy would require considerable time, perhaps years, zeroing in on the specific gene was a highly desirable goal in that it was a crucial step toward the big prize: to learn *how* the gene produced homosexuality.

Hamer showed his awareness of his data's ramifications by a cautionary note at the end of his paper: "We believe it would be fundamentally unethical to use such information to assess or alter a person's current or future sexual orientation, either heterosexual or homosexual, or other normal attributes of human behavior. Rather, scientists, educators, policy-makers, and the public should work together to ensure that such research is used to benefit all members of society." While such declarations of humanitarian concern were becoming commonplace in behavioral genetics papers, Hamer further demonstrated his concern by taking out a patent on the genetic region he had located, a move aimed at giving him a degree of control over future use of the information.

Most gay organizations heralded the finding, as they had Simon LeVay's discovery two years earlier, deeming both studies as substantiating what they had long believed, that homosexuality was not a choice but innate. A few gay groups protested the data, claiming it would be used to screen for gayness, to wrongly stigmatize people who might not be affected, and to attempt curing those who were. I watched astounded as Simon LeVay was attacked on a TV talk show by a gay man incensed at the suggestion his homosexuality was not pure choice. I suspect most homosexuals were as surprised as I was at this position.

Interesting as the arguments were, they demonstrated the speed with which discussion of behavioral genetics developments shifted from whether or not a finding was valid to whether or not it was desirable. For many, the latter controversy is irrelevant to any scientific insight—but especially to Hamer's, which most people saw as an important leap ahead in unraveling a complex and puzzling condition.

WHILE DEPRESSION AND HOMOSEXUALITY were two of the most interesting conditions to the behavioral geneticists, other traits were also being investigated by molecular biologists. Traditionally, I.Q. headed the list of traits about which curiosity ran high, but that dicey subject had all but been driven underground in 1969 by the outcry caused by Arthur Jensen's painful conclusions about racial differences. It was not to stay underground, however, and reerupted with full fury in late 1994 with the publication of Charles Murray and Richard Herrn-stein's book *The Bell Curve.*

Up until this revival there seemed to have been a tacit compact among those who knew the score, so to speak: Educators could continue using I.Q. measures for appraising their students *provided* that psychologists would shut up about group I.Q. differences. (*The Bell Curve* and the revival of the I.Q. wars will be discussed at the end of the next chapter.) Once the Jensen explosion had subsided into a period of peace, the lull permitted molecular biologists, without loss of political virtue, to join in the quest for the behavioral genes to which twin and adoption studies had pointed.

The next important advance occurred in Holland in 1993. Han Brunner, a geneticist at the University of Nijmegen, published a paper in *Science* about a Dutch family with a history among its male members of bizarrely aggressive, often violent behavior. A large number of males in the family reacted to minor frustrations and stress with wild outbursts—shouting, swearing, and sometimes assaulting the person they saw as their problem. The aggressive spasms included exhibitionism, arson, and attempted rape. Other males in the family were similarly afflicted, but to lesser degrees, perhaps only shouting and using obscene language. Still other males in the family showed no unusual behavior of any sort.

In the afflicted males, Brunner claimed to have found a small defect in a gene that produces an enzyme, monoamine oxyidase-a, essential in the breakdown of the chemicals that enable communication between brain cells. While some of the mechanisms are still a mystery, the Dutch researchers believed that because of the absence of the critical enzyme, those with the faulty gene accumulated excessive deposits of powerful neurotransmitters like serotonin, noradrenaline, and dopamine. These buildups led to the explosions of violence.

If Brunner's claim held up, it would be the first time a specific gene had been found that was responsible for a specific form of behavior. But even more exciting to some was that Brunner also believed he had a good idea of the way the gene functioned to produce the violence. If his explanation proved valid, it would be a major step forward in unlocking the mysteries of the "black box."

While many saw the study as a milestone in behavioral genetics, the authors demonstrated the now obligatory caution in presenting their results. Their finding, they wrote, said nothing about violence in general but concerned only one form of violence, probably an extremely rare form. The wariness was picked up by the scientific media reporting the study. While most publications played up the study in prominent stories, they watered down the implications until the public was left with the ludicrous picture of scientists having established a genetic link for the violent aggression of just one Dutch family, while all the other people in the world who threw plates at loved ones, made scenes in public, exposed themselves, and attempted rape remained as free of genetic influence as they had been in Skinner and Wilson's behaviorist cosmos. None of the writing about the study gave any reason to believe that other violent people had anything wrong with their genes but were violent merely as a logical reaction to their infuriating environments.

One rowdy Dutch family does not a scientific revolution make. But one Dutch family might explain the hyperviolence of countless people, alive and dead, who were never studied in this way. Even more important, one Dutch family can show the way in which genes control neurotransmitters, which control behavior. For all the modesty and caution of Brunner and his group, most observers knew their discovery was very important.

There was another reason for this caution. The possibility of a link between genes and violence is almost as politically charged as claims of a link between genes and I.Q. Some see any discussion of the topic as a sneak attack on America's black population, which, at the present time, is responsible for a disproportionate amount of the nation's violence. Although not broken down along racial lines, evidence exists of a genetic basis for at least some forms of violence. This information and its potential usefulness for society will be discussed in chapter 17.

THE DUTCH VIOLENCE STUDY gave a much needed boost to the morale of the Lohengrins of behavioral genetics, who had, on several occasions, thought they had found their Holy Grail, only to have it prove an alluring fake. Even the Dutch study's results, persuasive as they were, lacked replication by others, leaving the well-publicized failures looming over the field. Most behavioral geneticists, however, saw the failures as temporary misfires in a contest in which bull's-eyes were only a few volleys away.

In addition to the damage to morale, the premature cries of "eureka" had the inevitable effect of making everyone in the field exceptionally slow to claim success. This was particularly apparent in 1994, when a group researching manic depression under Wade Berrettini, a psychiatrist at Jefferson Hospital in Philadelphia, was certain they had found a gene that in their guarded words, "increases the susceptibility to bipolar illness." This time the depression-gene "sighting" was no place near the locations pegged by the Amish and Israeli studies, but on a different chromosome altogether, number 18.

While admirably muted in their announcement, the group's finding seemed to herald a game of follow the bouncing genes—not to be confused with Barbara McClintock's hopping genes—and maintained to a degree the shooting-from-the-hip impression gene hunters were making on outsiders. Of course, because the genome is such a vast molecular configuration, for a broadly defined disorder like depression, there is no reason why all three locations couldn't be correct—each for a different type of depression or each an ingredient in a depression that stemmed from a combination of genetic anomalies.

Unlike the earlier behavioral-gene studies that had been trumpeted from newspaper front pages, the Berrettini study was announced in a discreet throat-clearing in the July 1994 issue of *Proceedings of the Academy of Natural Sciences*. The project had been begun ten years earlier, so had been in progress at the same time as the Amish and Israeli studies; its team of researchers was all too aware of the embarrassment that can come from premature announcements. Its first six years were spent locating a number of afflicted families, enough for a meaningful study. Eventually the team found twenty-two families, each averaging about seventeen members and, each with approximately seven or eight cases of depression, for a total of 365 individuals. (One family had been "borrowed" from the Amish study, which was a nice instance of scientific cooperation.)

Starting with no prior assumptions and no clues such as a probable X chromosome linkage, the Berrettini group adopted a strategy of searching the entire genome, a task so colossal as to have been inconceivable a few years earlier. As if in punishment for their ambition, the group was as unlucky as the Huntington's people had been lucky in their choice of the area of DNA to start their search. Three years passed before the Berrettini group found the anomaly they were seeking on chromosome 18, one of the last places they looked.

When I spoke with Dr. Berrettini, he downplayed his find. "I wouldn't call if a breakthrough," he said. When I asked him, however, if his research had succeeded where the Amish study had failed, he was a bit more upbeat. "I hope so," he said, then added, "The gold standard for this sort of research is whether or not independent confirmation can be observed. That is, can you collect a second series of families and see the same results? I understand this has now been done."

WITH TWO FIRM VICTORIES in establishing a genes-behavior link on the DNA itself, few human geneticists doubted that others would be coming rapidly. By the summer of 1994, the process appeared so solidly on track that attention was jumping ahead to the next step in enlightenment: the mechanism by which gene expression translated into behavior. Much of the process appeared to involve gene products like hormones, enzymes, and neurotransmiters, as Jerome Kagan, Brunner, and others were discovering. One hormone in particular, testosterone, was receiving considerable attention, in part because of its relevance to the hot topic of gender but also because it was proving so interesting in itself. In his 1994 book *Social Structure and Testosterone*, Theodore Kemper of Rutgers University summarized the extensive research on testosterone and its influence on behavior. His central thesis went beyond the increasingly established effect the body's internal chemistry has on human behavior but rather examined the surprising effect the environment has on body chemistry.

A noteworthy example Kemper offered involves leadership. A number of studies had shown that testosterone levels are higher in men in leadership positions than in subordinates. On hearing just this part of the findings, few would sense anything remarkable about it, assuming the higher testosterone level was a reason, perhaps *the* reason, that the men were leaders. As logical and pro-intuitive as such an assumption might seem, it would be wrong. It was later discovered that the testosterone level rose *after* the men became leaders (which, for the ambitious, may well be an important secret of the universe). The added testosterone was an effect, not a cause. Similar environmentally induced fluctuations have been noted as well with serotonin. At a time when psychologists were discovering that the environment has less effect on personality development than previously thought, biochemists were coming along and saying the environment can have profound effects on internal biochemistry, which in turn affects personality.

The responsiveness of testosterone to life situations was demonstrated in studies involving different environmental situations. One of the more interesting found that U.S. Naval Academy midshipmen had far lower testosterone levels in their first year, when they were miserable plebes, than when they reached the exalted upper grades. An added curiosity that turned up in an unrelated study was that after having sex with a partner, a male's testosterone level rose, but after the same men masturbated alone, it *remained the same*. This would suggest that testosterone is not only highly responsive to surrounding situations but also that it is not easily fooled as to whether it has just experienced real turtle soup or merely the mock.

All of this had big implications for the genetics of behavior. The responsiveness of gene-produced testosterone to environmental conditions and the known effects higher levels of testosterone have on behavior raises large questions about the gene-environment interactions. Just as it is no longer possible to think of the environment as a monolithic and all-powerful molder of personality and behavior, so does the testosterone research suggest ways in which the environment, even within a normal range (*pace*, Sandra Scarr), can perhaps affect genetic expression.

Other research is turning up entirely new complications to the gene-enviroment equation. Stephen Suomi of the National Institute for Child Health and Human Development, in experiments with rhesus monkeys, found that when infant monkeys who had been bred to be inhibited and fearful were given to foster-mother monkeys who had opposite personalities—that is, uninhibited and fearless—not only did the infants become bold like their new mothers but they also developed the relevant brain chemistry, which for boldness meant low levels of norepinephrine. Most surprising of all, the changes appeared to be permanent.

This field of research into environmental alterations of hormone action, and in some cases gene action, is one of the last frontiers of the entire biology-environment conundrum. Other experiments indicated that repeated stressful experiences can turn laboratory animals into nervous wrecks, which is not so surprising. What was surprising, however, was that the prolonged stress-exposure brought about *permanent* changes in their nervous systems, including the expression of some genes. According to Suomi, "experience can push genetic constitution around. Its effect is so profound, I'd call it temperament." This doesn't overturn existing behavioral genetics thinking in that repeated stress falls outside Scarr and Plomin's "normal range," but it still bothers those who like their genes-environment dichotomies clear-cut.

THE DYNAMIC COMPLEXITY of norepinephrine may create an unnecessarily complicated example of the mysteries that lie between a gene and a behavioral trait. A clearer, more manageable example turned up late in 1993 in experiments done on an unprepossessing creature, small and furry, the prairie vole. In the latter half of the century, animal-behavior studies have contributed mightily to the revolution in the understanding of human behavior. Numerous research projects have pried open the gene-rooted behaviors and social organization of many species and, with similarities to human behavior that were too obvious to deny, played a major role in the seismic shift in thinking about human nature over the past forty years. While this science of ethology has increased understanding of such human activities as mating, political organization, and altruistic acts, recent research on this one small mammal, the vole, is highly suggestive of the way gene-produced body chemicals lead to specific forms of behavior.

Experiments on voles done at the National Institutes of Health by Thomas Insel found that two hormones present in voles as well as in humans, oxytocin and vasopressin, had important effects on the animals' approach to mating and to family life, aspects of behavior that are very distinctive in voles (and endlessly interesting to humans). The two hormones were long known to affect, in both voles and humans, physiological functions such as blood pressure and, in females, milk production. But by injecting the voles with a drug that blocked the hormones' effects, Insel found not only the expected physiological changes but also marked *behavioral* changes as well.

Voles, like many mammals and especially rodents, have rigid mating and parenting habits. Both sexes are monogamous, for example, and as parents are obsessively doting. When the two hormones were blocked, however, dramatic changes occurred in both of these characteristics. Generally, when male voles mate for the first time, they immediately become superattentive to the female of choice and fiercely protective of her—model husbands, in short. On the other hand, when vasopressin was blocked in the males, they remained as keen on sex but after enjoying it showed a caddish lack of interest in their partner. Even worse, if other males approached the females with carnal intentions, the husbands shrugged and wandered off.

A variation on this experiment was tried with results reminiscent of Tristan and Isolde's love potion. When a blocked male vole was given vasopressin treatments, he fell in love with the first female at hand, even before the usual coital trigger to affections. A big dose of vasopressin was enough to make him as ardent and protective as he would have been had he mounted her. With the hormone you saw the usual vole behavior, devotion with respect. Without it, you saw callous indifference.

The other hormone, oxytocin, appears to have as strong an influence on female voles' parental instinct as vasopressin does on males' mating behavior. For several years it has been known that the hormone makes females of a number of species eager to mate, keen on cuddling, and tireless in caring for their young. This raises the possibility that the reason human mothers submit to endless drudgery of rearing their young may not be because of parenting theories, religious upbringing, family values, or memories of their own childhoods. All such environmental elements may reinforce the self-sacrificing maternal behavior, but the real motivator may be a protein manufactured in the brain over which mothers have no control.

Curiously, other vole genuses lack these reactions to alterations in the levels of the two hormones, but evidence is accumulating that they may produce similar effects in humans. If this should turn out to be true, the commercial possibilities are limitless. Mothers who are considering abandoning their young could pop an oxytocin pick-me-up and return happily to the child-rearing grind. Or child-welfare agencies could require administering an occasional oxytocin injection before handing over cash benefits. When welfare mothers suspect their indifferent husbands are about to skip out on them, they could drop some vasopressin into their coffee and find themselves with story-book lover-dads. Yet another application would be the vasopressin nightcap, which would guarantee women that casual male pickups would send flowers or at least phone the next day.

TWELVE

THE UPS AND DOWNS
OF HUMAN NATURE

THE NATURE-NURTURE DEBATE has been raging, in one form or another, as long as humans have reflected about themselves. Adam and Eve started things off by calling attention to their innate nature (good but weak), while their nemeses, the serpent and the apple tree, could be considered environmental influences—leaving only the question of whether the Garden of Eden, as a rearing environment, fell within Sandra Scarr's "normal range." There have been so many fluctuations in the fortunes of the biological versus the environmental positions that it was inevitable that in recent times apathy has set in among bystanders. First one side's up, then the other; exhausted by these pendulum swings, much of the public has developed a cry-wolf indifference to the issue. That this reaction should happen now is both ironic and unfortunate in that science finally appears to have resolved the problem once and for all.

The conclusion that both sides now seem to agree on is that nature-or-nurture is a nonquestion. There was nothing wrong with the two concepts—nature, for genes and biology; nurture for the environment. There was plenty wrong, it turns out, with the "or." The most cogent dismissal of this thinking must be that of Martin Daly and Margo Wilson, who wrote in their landmark 1988 book *Homicide*: "One might just as well ask whether hemoglobin or air is more essential to human survival." This is a particular apt analogy, not only because *both* genes and environment are fundamental to development but also because their interaction is too.

The question, then, turns out to have been bogus, at least in the either-or form in which it was cast. Still, the longevity and virulence of the debate is surely a result of its relationship to fundamental questions on man's nature that scientists and thinkers have grappled with throughout history. And because the conclusions about humans were so momentous, the matter has been the staked-out turf of powerful political and religious institutions whose systems were erected upon their conceptions of this essence, of human nature itself.

So the debate has not been an abstract conundrum, a pastime for medieval monks, dormitory bull sessions, or competing schools of psychology. It is a question that can determine the way in which humans are reared, governed, educated, punished—and how they see themselves in relationship to their gods, to their governments, and to one another. It is no wonder that throughout history entrenched powers, spiritual and temporal, have ferociously defended their dogmatic pronouncements on mankind's essential nature—and many still do.

In looking over this history of fluctuating views about human personality and behavior, one of the most revealing areas is the changing beliefs about mental illness over the centuries. This specialized area might sound extraneous to discussions of normal behavior, but it is closely related, in that theories about mental illness invariably rest on theories about normal brain function. Because for much of this century, non-ill behavior was not seen as having any direct connection to our physical selves (mind and body were distinct entities), the approaches to mental illness provide a rare glimpse, basically the *only* glimpse, of the evolving view over the centuries of the mind-body connection and earlier attempts to explain individual differences.

The ancient Greeks had a simple, straightforward view: If a person was mentally ill, it was a result of a physical malfunction of the brain. According to Hippocrates, insanity was brought on by an imbalance in the four basic body fluids, or humors—blood, phlegm, yellow bile, and black bile. This analysis, which has long appeared more metaphorical than scientific, now seems to be closer to the truth—Jerome Kagan's neurological broth, for instance—than the insanity theories that held sway for the two millennia following Hippocrates.

For many of these centuries, lunacy was explained as possession by witches or evil spirits. The condition was considered a punishment for some terrible transgression, an explanation that conveniently justified killing the afflicted or in other ways disposing of them. In the Middle Ages, this folk wisdom was given theological respectability by the Church, which held the devil himself responsible for crazed behavior. Throughout this period, more thoughtful and skeptical people toyed with other, less fanciful and more evidence-based explanations; but scientific theories that conflicted with Church teaching invariably ran into trouble. With the perplexing phenomenon of insanity, few scientists were willing to go up against the Church armed with little more than hunches. In fact, for centuries there weren't many hunches. Baffled by insanity, most people were content to shake their heads and accept the Church's dictum that such behavioral anomalies were God's will or the devil's mischief.

In the eighteenth century this resignation began to weaken as more and more doctors sought medical explanations for mental problems. Prominent among these was the eminent Philadelphia physician Benjamin Rush, a signer of the Declaration of Independence, who diverted his political rebelliousness to the prevailing beliefs about many diseases, in particular, insanity. Rush was convinced it was caused not by demons and devils, nor by bad behavior, but by a physical malfunctioning of the brain.

Throughout the nineteenth century a handful of physicians pursued this concept. In 1844 another American doctor, A. L. Wigan, wrote a book entitled *A New View of Insanity: The Duality of the Mind—Proved by the Structure, Functions and Diseases of the Brain*. While Dr. Wigan did not make a significant advance in the treatment of mental illness, he was one of the growing number of scientists pursuing the concept that mental function resulted from the brain's physical makeup, not from witches, spirits, or God's will. This area of investigation was also being pursued in Europe and culminated in the science of psychiatry, a term that included *all* therapies aimed at curing mental disorders.

Toward the end of the century in both America and Europe, the physiological approach dominated the field. One of its leading proponents was the German psychiatrist Emil Kraepelin, who made a major contribution to the newborn discipline by classifying psychosis into two main categories: dementia praecox (later renamed schizophrenia) and manic-depressive psychosis. Kraepelin was also the first to use objective tests and measurements to study drug effects and mental disorders. As his influence in Europe grew, Kraepelin's physiological approach to brain function appeared to be the path of the future. Instead, it was swept aside by the advent of Sigmund Freud and his revolutionary psychodynamic vision of mental illness and of personality itself.

IN THE CENTURIES before Freud, the great philosophers occasionally turned their attention to the determinants of behavior; but they did so from such different perspectives that it was hard to realize they were talking about the same thing. Descartes, with his theory of involuntary and voluntary behavior, postulated the concept of internal forces or drives within the human that interacted and often struggled with voluntary behavior; this model of mindless impulses in conflict with conscious thought is not far from current thinking about the genetic influence on behavior. Rousseau believed that man was born better than a blank slate; he was born good, self-reliant, and free, but had been corrupted by an evil society. Yes, humans had innate dispositions, said Rousseau, but they were no match for the world's power to pervert.

In the nineteenth century, Schopenhauer, in his major work *The World as Will and Idea*, defined his concept of "will" as, among other things, a mysterious, self-generating force that drives the world. It was not a force imposed from without but something that existed in everyone. As a key statement put it, "The world as the thing in itself is a great will which does not know what it wants; for it does not know but merely wills, simply because it is a will and nothing else." A geneticist, particularly one of the Richard Dawkins "selfish gene" school, could be forgiven for giving a genetic interpretation to Schopenhauer's concept of an unthinking force that drives the world.

Prominent in the philosophical camp that leaned more toward external influences on behavior was John Locke, who argued that all knowledge, including behavior, was derived from observation of the external world, *not* from innate sources. John Stuart Mill elaborated on this view, claiming that sensation and experience were the elements that molded human nature. In these and other systems, behavior was something instilled into humans from the outside—from, you guessed it, the environment.

In spite of these somewhat fanciful stabs at understanding the engines of behavior, the subject of behavior was, for these philosophers, some distance from their central concerns: Man's place in the universe, his relationship to God, the nature of knowledge—these were the big issues. Why the majority of humans behaved as they did was of secondary interest, if of interest at all. Man was man, woman was woman; dramatic variations from human to human were always fascinating—from Lucretia Borgia to Michael Jackson—but the intricate array of traits we all have in common were taken for granted and ignored.

Earlier in the nineteenth century, just before Freud derailed the scientific trend toward a physiological view of mental function, two scientific developments occurred that would gradually have enormous impact on the understanding of human behavior. In light of the monumental import of these intellectual events, recognized by many at the time, it is remarkable that their full relevance for humans was not realized for close to a century. One was the publication in 1859 of Charles Darwin's *On the Origin of Species*, which launched his theory of evolution that now informs all reputable thinking about life on earth, including human life. The other great development was Mendel's discovery of the laws of genetics, which he published in 1866.

Darwin's huge idea was quickly picked up by prominent intellects of his day, some of whom immediately set about exploring its implications, as did Darwin himself for the remainder of his life. Still, the ramifications were so vast, and so threatening to man's lofty self-view, that the implications for human behavior were either set aside or ridiculed. The most famous example of the derisive reception was when Bishop Wilburforce, a leader of the anti-Darwin forces, regaled a meeting of the Royal Academy of Science when he asked Thomas Huxley, Darwin's chief proponent, if his ape forebear were on his father's or his mother's side. Many joined in on the jibing. In Gilbert and Sullivan's *Princess Ida* the man-hater Psyche sings of an ape in love with a woman:

> He bought white ties, and he bought dress suits,
> He crammed his feet into bright tight boots—

And to start in life on a brand-new plan,
He christened himself Darwinian Man!
But it would not do,
His scheme fell through . . .

Even though Darwin was quite explicit about his belief that humans were just another animal, the idea met with far greater resistance than the more earth-shaking theory of natural selection. Many could go along with the idea that all species descended from a common ancestor, even man. But the idea that man only differed from other species by a few degrees of intellect or civilization was anathema to many. Our species may have evolved from the apes, the thinking went, but we evolved into something profoundly different, so different in fact, we have severed all behavioral ties with the animal world and now operate purely on free will and brain power.

It was not only the religious and the species-proud who refused to follow Darwin all the way to his logical conclusion that humans inherited behavioral traits just as animals did. Many prominent scientists could not swallow this aspect of Darwin's system—even Alfred Wallace, who arrived at the theory of evolution by natural selection at the same time as Darwin and who might have been expected to be an unwavering supporter of his cotheorist. To Darwin's dismay, Wallace deviated on this one major point, proclaiming, on spiritual grounds, that man was a special creation, separate and distinct from all other species.

The mere title of Darwin's second major work, *The Descent of Man*, points up his belief that the entire theory of evolution, while explaining all life on earth, culminated in a fresh understanding of the human. On the other hand, Wallace and some others were saying to Darwin, we concur with your entire complex theory, we agree that all species descended from the same organism, but man, who may have evolved the same way, has been divinely endowed with unique qualities that remove him from this evolutionary choo-choo train. For Darwin, these defections were a major disappointment. After convincing large segments of the intelligentsia that all living things had not been produced in a few days of God's creative week but rather evolved over millions of years through a natural process, Darwin was confronted by allies who balked at his all-important conclusion about humans.

The specialness these holdouts awarded their species applied particularly to behavior. Arms and ears might have evolved, but human behavior was the product of undiluted reason, pure cognition. That lower life-forms were observed to have inherited automatic behaviors or instincts was not seen as having any relevance to human behavior. Their behaviors were immediately dubbed instinctive for the circular reason that being lower life-forms, they could not think. This conclusion was as little justified as the conclusion that we humans thought about everything we did. By imposing simian ancestors on humans, Darwin had inflicted considerable damage to his species' dignity and self-esteem. Consciousness and reason-driven behavior were clung to as essential dividers between humans and all other creatures, bulwarks to protect our unique place in creation.

In spite of the triumph of Darwinism, rather than Wallacism, a dogged resistance to man's genetic affinity to other animals resulted in an invisible intellectual wall going up between man and the rest of creation. It was as though the thinkers and opinion-makers were saying, "We will grant you your hard-to-swallow evolution *provided* you don't include us." It was all right to place physical man in the system (our bodies and moving parts), but our greatest glories—our minds, our behavior, our aspirations—such wonders cannot be explained by evolution; they must come from somewhere else, probably a God who was partial to humans. The implicit deal was similar to the Continental Congress's permitting southerners to keep their hateful slavery in order to pass the Constitution, a compromise that prompted Abigail Adams to refer to slavery as the Revolution's "unfinished business." It appeared that a full recognition of modern man's evolutionary heritage—body and behavior—was, until recently, the unfinished business of the Darwinian revolution.

Although Darwin himself, unlike the Continental Congress, was not willing to make any such concession, the deal-makers had their way, not by jettisoning aspects of Darwin's scheme or demanding a scientific line-item veto, but by simply warning others to be very careful in extracting inferences about humans from Darwin's system. "Very careful" translated to "make no inferences at all." Although a damaging compromise, this tacit compact held for nearly a century. Any scientists intrepid enough to extrapolate theories about humans from animal-behavior studies were generally derided as anthropomorphic ninnies, passing off as serious science fantasies along the lines of the film *Babe*, with its barnful of chatty animals. The prevailing wisdom came to be that animal studies were interesting but of little relevance to us oh-so-special humans, a species who had gloriously transcended, broken free of, triumphed over, evolution.

Maintaining our specialness hasn't always been easy. Over the years the effort has produced a catalog of characteristics said to be unique to humans—the opposable thumb, toolmaking, self-awareness, a moral sense—all of which were demolished by one annoying species or another that was found to possess the same ability. One of the silliest was the claim by a group of anthropologists that humans are the only species to laugh. They were shortly confronted with the discovery of a Central American howler monkey who spends its days in the top of trees, laughing its head off—at human arrogance most likely.

That it would also require a hundred years for awareness of the application of Mendel's laws to human behavior is less surprising. Unlike Darwin's theory, which rocked the world, Mendel's equally monumental findings were virtually ignored at the time and continued to be ignored for thirty years. It was not until 1900, sixteen years after Mendel's death, that a group of German scientists working on similar breeding experiments came upon his writings and credited him with the discovery of the basic laws of genetic transmission. But even when the importance of Mendel's system was recognized, only an eccentric few felt it had any relevance to human behavior; for nearly two-thirds of the twentieth century, almost none of the serious thinkers on the subject believed our behavior was in any way inherited.

If anyone could have brought home the connection, it was Darwin. A great misfortune of Mendel's belated recognition was that Darwin appears to have been unaware of his discoveries, despite the fundamental connection between the two theories. Mendel had, in fact, answered questions raised by Darwin's theory of natural selection—the mechanisms of inheritance—questions Darwin continued to grapple with for the remainder of his life. Mendel's first paper on his theories was published only seven years after *On the Origin of Species*. Even though the paper was sent to scientists and libraries throughout Europe and in America, it was ignored. For another sixteen years, Darwin would live and pursue his scientific investigations. There can be little doubt he would have been inspired by Mendel's discoveries of trait transmission. A poignant underscoring of the lack of scientific communication, if not the impotence of obscurity, was that after Darwin's death, a copy of Mendel's paper was found in his library, its pages uncut.

EVEN WITH THE PHYSIOLOGICAL APPROACH to mental illness of Kraepelin and other turn-of-the-century psychiatrists and neurologists, few of them would have seen any connection between the earthshaking theories of Darwin and Mendel and their own explorations into the determinants of human behavior. This intellectual separation between genetic inheritance and behavior was broadened greatly by the Freudian *Anschluss*, which put all emphasis on postnatal events in explaining behavioral anomalies. Freud never denied the existence of innate drives and instincts, although this was an area he found troublesome and on which he revised his thinking a number of times. In his system such inherited traits tended to be human constants (sexuality, self-preservation, and so on), while he explained behavioral differences between individuals, particularly the neuroses and psychoses that were his concerns, in terms of early experience that disrupted the normal development of these human givens and left in the mind—the subconscious mind—a residue of problem-causing influences.

This was an entirely new view of the way the mind governs behavior, although, as has often been pointed out, some of Freud's most important ideas had precedents in the writings of thinkers like Nietzsche and Schopenhauer. But never before had such powerful notions as repression and the subconscious been cast in a medical, therapeutic context. Since Freud applied his insights to all forms of aberrant behavior, from severe psychosis to mild neurosis, his theories had the effect of bridging the gulf between insanity and normal behavior. Increasingly, human behavior would be seen as a broadly varied continuum, and in Freud's view (and even more in the view of his disciples) the variations sprang primarily from the experiences of childhood or early infancy and the resulting imprinting on the ever-active subconscious. The happy corollary was that if problems resulted from mishaps along the developmental path—denting the developing child, so to speak—the damage could later be rooted out to restore normalcy. Freud's exciting theories had a major impact in Europe, even more in

America. As a result, the notion of biological and genetic influences on behavior was swept aside.

With the disposal of biology as a behavioral influence, the great intellects of the early twentieth century had created a vast playground for arcane and fanciful constructs of cause and effect. It was far more stimulating and, in the end, optimistic, to lay neurosis and insanity at the feet of a snarl in the parent-child dynamic—scenarios involving lush dramatic tools such as jealousy, fear, guilt, and so on—rather than to trace the malfunctions to a few wayward molecules. When such creative and visionary geniuses as Carl Jung and Otto Rank hooked their man-made systems of inner conflict and striving to the spiritual, they brought forth even more dazzling literary outpourings—from themselves and from gifted onlookers like Aldous Huxley and Laurens van der Post. Other sciences aimed at the human took the same giddy position: Human personality was a human construct. Flaws in construction could be corrected and humans could be built better in the future.

Despite the triumph of Freud's psychodynamic view of human behavior, Kraepelin and others in Europe continued to explore the physiology of brain function and malfunction. Psychiatric historian Nancy Andreasen, in her 1984 book *The Broken Brain: The Biological Revolution in Psychiatry*, points out that Freud himself had originally been drawn to neurology and the brain's physical malfunctioning, but that his Jewishness had barred him from the university clinics in Vienna that were essential for research in this field. Because his marriage in 1886 necessitated a regular income, he dropped his primary interest and began seeing patients at home. This unwanted clinical practice led to the theories that changed the world. While Andreasen's home-office explanation of Freudianism's birth rests on a number of large assumptions, it is still irresistible to conjecture that had Freud not been discriminated against, he might have followed the neurophysical path of Kraepelin and Alzheimer, whose biophysical view of mental illness Freudian analysis would all but bury for three-quarters of a century, and which has only recently returned to dominate the psychiatric mainstream.

After Freud's sole visit to the United States in 1910 for his famous lectures at Clark University in Worcester, Massachusetts (whose honorary doctorate of laws degree I spotted proudly displayed in Freud's Vienna waiting room on a visit in 1994), his fame and influence spread rapidly across the country, so that by the late 1930s almost every major psychiatric post in America—department heads at leading universities and the editors of the top professional journals—were committed Freudians. So complete was the takeover that most Americans thought that the terms *psychiatrist* and *psychoanalyst* were synonymous and were unaware that Freudian analysis was merely one school of psychiatry.

In Europe, however, Freudianism never achieved the domination it achieved in America. From the start, European scientists and neurologists were intrigued by Freud's thinking but never embraced it with the enthusiastic abandon of Americans, who increasingly saw the theories as the answer to everything worth knowing about humans. The phenomenon conjures a picture of Freud as history's most brilliant snake-oil salesman working a nation of gullible rubes. More likely the explanation lies in American optimism and hunger for new ideas. When the tide eventually swung back to a view of behavioral aberrations that included physical malfunctions and chemical imbalances, the collapse of Freudianism caused less upheaval in Europe than it has in America.

The fifty-year triumph of Freudianism in America was buoyed by the massive political tides that were welling up on both sides of the Atlantic. Virulent social injustice that had swept in with the post–Civil War industrial boom was feeding the growth of progressive movements such as socialism, communism, and other system-changing ideologies. All were aimed at building more just societies and all were based on a belief in man's capacity for improvement. Freud's ideas about the malleability of personality jibed neatly with these political visions. Opposing theories about innate, biology-rooted traits, with connotations of immutability, were incompatible with them.

It is possible that the good-news/bad-news stigma that would cling to the nature-nurture debate of later decades received a considerable boost from Freud's disciples, men and women heady with the prospect of curing mental illness and who didn't want to hear about stubborn genetic flaws. Snarled and mislearned behavior could be unsnarled and unlearned on the psychoanalytic couch. Inherited personality traits, if they existed, appeared as alterable as eye color and foot size.

Nursery mishaps could be remedied; congenital mishaps could not. The Freudians didn't so much deny the existence of innate traits and tendencies as they ignored them.

AT THE TURN of the century the Freudians were not the only group intoxicated by exciting new ideas about improving humanity. Some years earlier, Darwin's cousin, scientist and thinker Sir Francis Galton, who dabbled brilliantly in many areas, including meteorology and anthropology, pondered the ramifications of Darwinism for the future of the species. As a result of these ruminations, Galton launched eugenics, a movement aimed at encouraging the propagation of the most fit and discouraging the propagation of the indisputably unfit, such as lunatics and the deformed. Although the eugenics movement, which would later acquire an infamous name, was attributed to Galton, Darwin also pointed out that man was the only species with the ability and desire to overrule evolutionary pressures by helping the unfit. He worried, as did his cousin, about the long-term implications for the species of well-intended efforts to protect and care for "the imbecile, the maimed, and the sick."

Although the original purpose of eugenics was merely to reduce the numbers of genetic disasters born into the population and to encourage the "fittest" to have more offspring, the thinking was soon picked up and expanded by others who applied it to the horrendous social problems plaguing cities—poverty and the attendant crime and violence. Instead of blaming the economic system itself, the eugenicists blamed the unfortunates, who were seen as congenitally incapable of adapting to modern society. Since eugenics focused on inherited personality traits, its sorry and reckless career would for many years stigmatize behavioral-genetics thinking.

The movement took off rapidly and some of the day's most socially concerned and liberal intellectuals, such as George Bernard Shaw, Beatrice and Sydney Webb, and Harvard president Charles W. Eliot, embraced eugenics wholeheartedly. The eminent Harvard zoologist Charles B. Davenport launched a eugenics society in the United States that drew large numbers of adherents. An officer of Davenport's Eugenics Records Office on Long Island was Alexander Graham Bell, who had worked for many years with the deaf and had written papers about the hereditary basis of deafness. At the first International Conference of Eugenics, which was held in London in 1912, Winston Churchill was an officer. The movement was not just a fin de siècle fad, but flourished for many years. Davenport remained director of the Eugenics Records Office until 1934, and the theory was considered a respectable program for social betterment until the Nazis so grotesquely demonstrated the perils in eugenic thinking.

It is surprising that these early eugenicists failed to foresee the dangers in such procreational tinkering. In fairness to them, for all their tendencies toward paternalistic elitism, they were mostly decent types who were unable to envision that a member of their own species might one day apply their concepts in a drive to eliminate disliked races and ethnic groups. The monstrosities of Hitler demonstrated that when dealing with populations, there is a frightening proximity between the power to decide who is born and who is not born and the power to decide who lives and who dies.

Some people were strongly opposed to the eugenics movement long before Hitler. Their objections, however, were on ideological, rather than moral, grounds; for them, eugenics placed too much emphasis on genetic inheritance, with the implication that the only way to improve the species as a whole was to prevent the birth of the unfit. The theory implied that existing men and women were unimprovable. Still, for many years eugenicists and their focus on humanity's biological makeup were a highly influential group that grew steadily. For a long period they were the only group who proposed a way to solve society's most nagging problems, poverty and crime, without demolishing the social order.

IN AMERICA, the leader of the backlash to the eugenicists' burgeoning popularity was Franz Boas, an anthropologist from an affluent, educated German Jewish family, who had moved to the United States in the 1880s as a young man because he believed German anti-Semitism would hinder his academic career. Establishing himself at Columbia University, Boas eventually became one of America's most respected and compelling voices in the social sciences throughout the century's first decades. Early field research with Eskimos and Pacific Northwest Indians convinced Boas of the importance of culture in shaping personality and behavior. He did not dismiss the contribution of heredity, as did his followers, but he believed biological processes were separate and distinct from culture and could only be turned to for explanations if the cultural possibilities had been thoroughly explored and rejected. He also believed that culture was sui generis, that "culture came from culture."

Boas saw his mission as reversing the eugenics tide with all its implications of immutable human behavior that could only be controlled by regulating births. His references to inherited traits bordered on lip service; his emphasis was invariably on culture's massive role in determining human behavior. The encouraging corollary was that changing the culture could change the humans who are formed by it. For all his optimism, Boas was under no illusion this would be a simple matter. Indeed, he believed that man was "shackled by culture." Still, his worldview offered more workable, humane, and less drastic remedies for social problems than draconian efforts to regulate births. Populations didn't have to be culled for unfit parents; cultures could be changed to make them fit.

Boas's use of the emotionally charged term *shackled* provides a clue to the passion and liberationist's zeal of Columbia University's anthropology department in those pre–World War I years. Boas, and later his followers, were not just convinced they were right; they were convinced that eugenics and other schools of genetic determinism were dangerously wrong.

In the first years of the century, however, Boas was almost alone in his fight to halt the spread of eugenics among thinking people. But his gene-minimizing ideas had strong appeal among political liberals, who abounded in the country's academic and intellectual circles. Because Boas's culturalism struck hard at eugenic assumptions, political progressives—who had always hated the eugenicists' blaming poverty on the poor—were thrilled that such a respected scholar had provided an alternative vision for improving society. His thinking was eagerly adopted by many in the social sciences, and by 1920 Boas's cultural-determinist school was well on its way to becoming the reigning view of human nature.

With a missionary's conviction, Boas sought to inspire students with his vision. Two of these, Ruth Benedict and Margaret Mead, would surpass Boas in influence on the nation's intellectual zeitgeist as they beat back the forces of biological determinism and planted the flag of culturalism. Of the various forces that turned enlightened thinking away from biological determinism, these two anthropologists were among the most important.

In spite of Boas's success in winning adherents, his theory lacked powerful substantiation. To his mind, a "proof" would be the discovery of an existing human culture that was free of the behavioral problems that plagued American society. If such a group could be found, it would prove that humans were born free of behavioral leanings and were instead programmed by their culture to behave as they did. The underlying assumption, one that informs so much comparative anthropology, is that humans are pretty much the same genetically. If behavioral differences exist, they must come from outside the individual, that is to say, from the culture.

This is a hefty assumption and overlooks the possibility that Australian aborigines may differ from urbane Parisians or New Yorkers, not because of their culture, but because of minor genetic variations. There is a built-in circularity to the cultural determinists' argument that the existence of different cultures proves the human is a blank slate on which diverse cultures can write a wide variety of behaviors. Differing cultures can just as readily prove the opposite: that the slate is not blank, but the inherited writing on it is not always identical. Cultural variety may also indicate that the hard wiring shared by all humans is flexible and can be adapted to a large number of different expressions.

In the 1920s, however, anthropologists were exhilarated by the idea of a natural human, Rousseau's idealized man, a creature identical to us but untouched by the grindstone of Western civilization and therefore simple and good as we might have been and might be again. However shaky this premise, Margaret Mead made her famous trip to Samoa in 1925 eager to please her mentor by proving it was true. She would find what Boas needed to codify his theory, an idyllic culture so different from America's that it established once and for all the enormous malleability of human behavior. With so much riding on her expedition, it is not surprising that Mead believed she had found precisely such a culture.

Her landmark 1928 book on the expedition, *Coming of Age in Samoa,* depicts an idyllic society with values and mores diametrically opposed to those prevalent in America at that time. Among the long list of differences: Samoans were casual about sex (premarital as well as homosexual), indifferent to rank, devoid of passions, indulgent of their children, unpossessive of them, uncompetitive with one another, nonviolent, peaceable. The long list of attributes might have been a manifesto from the Woodstock Nation and stood as an eloquent rebuke to 1920s American society, which appeared to be destroying itself with opposite tendencies. Samoa sounded too good to be true. And, in fact, it was.

Mead's portrait stood for a half century as inspirational proof that culture was everything and inherited traits irrelevant, until Derek Freeman, in his exhaustively researched book *Margaret Mead and Samoa*, demonstrated how every one of Mead's observations about the Samoans was wrong. It was not a matter of her interpretation but of her facts. In order to debunk Mead's picture, Freeman used government records of Samoan warfare and uprisings, police records of violent crimes, and Samoan testimony about their preoccupation with rank and competition, their violent jealousies, their murderous obsession with chastity in unmarried girls. He also used first-person testimony of a large number of Europeans and Americans, combing detailed reports by explorers and missionaries who had been visiting Samoa for eighty years before Mead arrived. With this and other evidence Freeman showed that in virtually every instance the reality of Samoan life was the opposite of Mead's portrayal.

The picture Freeman paints of Mead herself carries an implied explanation of her upside-down account. He presents a twenty-three-year-old anthropologist, fresh from graduate school, on her own for the first time, eager to make a contribution to science—or at least to her mentor Franz Boas's anthropological vision, which was *her* science—exhilarated by her role as sympathetic scholar curious about the natives' uncorrupted ways, approving of everything she saw—and getting everything dead wrong in an almost Mr. Magoo kind of blind confusion.

The picture would be comic, ideal material for the late Charles Ludlam's Theatre of the Absurd, if it were not for this upended view of reality becoming the bedrock upon which much modern thinking about human nature was built. Mead's book was an immediate bestseller, continued to sell widely for many decades, and would eventually be read by more people than any anthropological book in history. Although no one, including Freeman, ever suspected Mead of deliberately cooking her data, the report is now seen by most scholars as a scientific embarrassment along the lines of the Piltdown man hoax. At the time, however, the enthusiastic reception of Mead's book marked the triumph throughout the nation's intellectual life of Boas's view of human nature as a nearly blank slate on which culture imposes personality and behavior. Till then this had been just one of many theories; Mead had made it triumphant fact with a delightful book about nubile virgins in a Gauguin paradise.

Freeman's debunking of Mead did not appear until 1983, fifty-five years after the publication of *Coming of Age in Samoa*. Throughout those decades there were a few boggles and grumbles from Samoans and others who knew the islands, but for the most part Mead's idyllic portrait of the wonderful Samoans stood unchallenged. At the same time, the book's major message, that humans can become whatever their culture cares to make them, became canonical truth upon which modern society could remake itself. Until her death in 1978, Mead remained perhaps the world's best-known and most influential anthropologist, an eminence she undoubtedly deserved because of her later contributions to the understanding of human cultures—and her much later acknowledgment of genetic components to behavior. She can hardly be blamed for the world's enthusiastic embrace of her youthful research debacle. In anthropology, as in show business, give the people what they want. . . .

When, in 1934, Mead's colleague and former teacher Ruth Benedict published her major work *Patterns of Culture*, it too made an enormous impact and reinforced the cultural determinism that Mead's book had validated. The Benedict book was translated into fourteen languages and over the next twenty-five years sold more than 800,000 copies. It was on course lists for every anthropology department in the United States. When I was at Yale in the early fifties, it was required reading for *all* undergraduates, an honor not shared by *On the Origin of Species* or *Moby-Dick*.

Benedict's book, which examines three disparate cultures, is far more accurate and scholarly than Mead's. It is also more forthright in its conclusions that human behavior is highly variable and a product of the surrounding culture. While Mead had let her portrayal of flaw-free Samoans stand as refutation of hereditarian views about human nature, Benedict was more confrontational toward those who still clung to notions of innate traits and tendencies.

It is remarkable to read Benedict's book today in the light of all the recent information about the gene-behavior dynamic. Even though the new data undermine her ultimate conclusion about the absence of inherited characteristics in humans, *Patterns of Culture* still emerges as a major scholarly achievement. Her research into three complicated cultures and her graceful and lucid presentation of her findings succeed admirably in the not minor task of establishing beyond question cultural relativism, the many workable variations in human societies. That she went on to offer this data as proving the absence of inherited behavioral traits appears almost a minor shortcoming when placed next to her brilliant analysis of dissimilar cultures. Unfortunately, it was the aspect of her book most enthusiastically seized upon by social thinkers and policy-makers.

Like others of the Boas school, Benedict was high on diversity and saw the many cultural variations as establishing the unlimited malleability of the human animal. In studying other cultures, today's evolutionary psychologists see something quite different; they focus instead on the vast number of similarities that run through every culture—incest taboos, altruism, and religiosity are three of the most frequently cited. The global presence of such traits—and of a far larger number of behavioral commonplaces—rendered them less interesting and, as a result, less noted than the differences, especially by anthropologists like Mead and Benedict, who were grooving to the differences. The ubiquitousness of specieswide traits made them invisible.

Now, however, such across-the-board similarities are cited as strong evidence of the species' shared genes. Travelers today complain of the growing similarity of the world's peoples, whose quaint differences had always been a lure to get up and go. One could speculate that the ease with which McDonald's, Coca-Cola, the film industry, and other multinationals are gratifying tastes shared by all humans is evidence of specieswide behavioral genes. Cultural variation may have been an aberration of geography and is rapidly being obliterated by telecommunications. Human sameness, for better or worse, is where it's at.

Benedict was among the few from her camp who gave serious attention to the customs and taboos found in virtually all cultures.

Admitting such planetwide behaviors presented problems for cultural determinists; Benedict explained them as traits that reach back to the species' origins, to a time when humans were one race, one culture, and few in number. They originated, not in our genes, she claims, but were adopted by the primal tribe and have been passed down to every descendant tribe that spread throughout the world. These traits have been with us so long, she believed, that they have become "automatic," which explains their survival in every culture in the world. She called such universal behavior "cradle traits" and refused to yield them to the genetic camp.

Yes, present-day Australian aborigines may have no cultural interchange with Swedish professors, but at one time a few hundred thousand years ago they were all part of the same tribe and certain tribal customs—not sleeping with your sister, for instance—have survived the eons and are cultural artifacts subscribed to by all humans. Her clever explanation seems an almost minor variance in interpretation from the behavioral-genes explanation, but it is one that produced a vastly different conclusion. In her model the human remained a blank slate at birth but a slate that had been scribbled on when it was fresh from the evolutionary factory. The scribbling was passed on through the ages culturally, not genetically. One weakness of Benedict's construct is that cultural traditions, which have repeatedly proved to be frail determinants of behavior, were subjected to eons of vicissitudes, exigencies, and human caprice, whereas the genes, should they underlie aspects of culture, were safely encased within each human cell and were passed down for the most part intact throughout the millennia.

The turn-of-the-century eugenicists, in their gusto for the new Darwinian thinking, all but ignored environmental influences and may have been guilty of establishing the ersatz either-or premise that has long dogged the culture-versus-inheritance debate. Boas's school of cultural determinists was no more balanced. In their counterattack, they went to similar extremes. When eugenics top dog Charles Davenport could write of "the fundamental fact that all men are created *bound* by their protoplasmic makeup . . . ," it is not surprising that Boas's disciples, in refuting such strident biological determinism, would go equally overboard in their sweeping claim that cultural was all. Both sides, fueled with a modicum of evidence in their favor, crammed human nature into their opposing ideological boxes.

AT THE SAME TIME Boas and his anthropological cohorts were winning the field for cultural determinism, reinforcements from another discipline, psychology, were arriving with a similar anti-inheritance system. In the 1920s John B. Watson launched the behaviorist school of psychology—later championed by B. F. Skinner—which denied the importance of inherited traits, claiming instead that the primary determinants of behavior were early learning and conditioning. The theory harmonized beautifully with the cultural determinists, not to mention the progressive political currents, so took hold and rapidly came to dominate American psychology.

Looking back at this period, Robert Plomin and his coauthors of *Behavioral Genetics: A Primer* summed it up by saying, ". . . the legacy of John Watson's behaviorism from the 1920s was the detaching of the study of behavior from the budding interest in heredity." Since behaviorist theory acknowledged culture as an important conditioner, the new psychology could align itself with the Boas group of cultural determinists. Further reinforced by the Freudians and their preoccupation with childhood experience, the three intellectual juggernauts were able to drive from the field those who believed in any degree of biological determinism. Psychologists could now say, "The environment is all powerful and if you don't believe us, just ask anthropologists and psychoanalysts."

By the 1930s the door was effectively closed on investigations into inheritance of behavior and personality. In the first two decades of the century, a number of eminent psychologists had studied instinctive behavior in animals, always with the hope of gaining insight into human behavior. In 1916 Ada Yerkes, working with albino and normal rats, carried out one of the first studies on the inheritance of learning capability. Other prominent psychologists, such as William McDougal and Edward C. Tolman, also among the early "rat-runners" (as they were affectionately dubbed), dedicated themselves to breeding experiments that threw light on the mechanisms by which behavioral traits were transmitted from generation to generation.

But the influential Watson, whose background was also in animal research, went to war against the entire instinct concept. Behavior in rats, mice, and humans might appear instinctive, he said, but in reality it resulted from *learning*, which in turn resulted from rewards and punishments. Behavior, therefore, was totally environmental, derived from the "culture"—whether human or animal. For a time the two opposing schools of psychological thought progressed side by side—conducting experiments, producing papers, holding meetings—and the whole subject could have settled into a stimulating scientific debate had not a third element intruded: political ideology. Sadly for the spirit of unbiased inquiry, both positions became identified with specific political outlooks, and this linkage wrenched the debate from science to ideology. The destructive effects of this counterproductive and mostly false association of ideas are still being felt to this day.

IN THE EARLY PART of this century, when the instinct-versus-learning argument thrived, political turmoil brought on by the Industrial Revolution was shaking the social and economic stratifications that had been in place for centuries in both Europe and America. Appalling worker conditions and rampant poverty (and the crime and violence that came with it) made clear the need for drastic reform, if not for altogether new social structures. Movements for fundamental change gained momentum, particularly among intellectuals and other socially aware men and women. By the 1930s the largest universities and most scholarly journals were dominated by those committed to these political programs, all lumped under the term *progressive*.

It soon became a basic tenet of progressive thought that the biological view of human nature was antiegalitarian, counterprogressive, and, in the view of some, nothing more than bogus science tricked out to justify the status quo. To substantiate the latter view, the progressives pointed to the Social Darwinism of Herbert Spencer, the English philosopher who seized on Darwin's discovery as proof of the inevitability of social stratifications. Spencer also believed that the theory of natural selection demonstrated the folly of governmental attempts at interfering with "the natural order."

Spencer's nasty spin on Darwinism was a comforting theory for rapacious capitalists in that it implied that their heartless exploitation of their fellow man was a manifestation of nature's design. Spencer's "Social Darwinism" was also agreeable news for the privileged classes, since it held that poor laborers—whose joyless lives might otherwise have inspired pity, perhaps assistance—had evolved to be poor. On the other hand, those whose main effort in life was collecting Chinese porcelains in the family's Belgrave Square mansion were so blessed because they had more of that fitness stuff, whatever that might be.

But robber barons and British aristocrats did not determine the prevailing intellectual wisdom. Writers and academics did, and within that group progressive thinking, including the doctrines of behaviorism and cultural determinism, ruled the day. While no one, then or now, doubted the sincerity of the scientists behind those doctrines, it now appears that their scientific judgments were strongly influenced by their utopian visions. The thinking seems to have been that in order to implement their ideal societies, they had to first disconnect human behavior from intractable biology.

In spite of worthy motives, their easy victory for rigid environmental determinism—now believed by almost no one—stands as a warning of the way in which erroneous science can be buoyed to broad acceptance if in tune with the political zeitgeist (which, of course, is the argument now used by the retreating environmentalists, who explain the successes of behavioral genetics with a swing toward conservatism). It also shows the ease with which valid science, if it appears to challenge the political temper, can be ignored or buried. The most flagrant example of this occurred in Soviet Russia, the country to which most progressives in the 1930s were looking for leadership in advanced social thinking. During the first decades of the century, Russian science had produced a number of the world's finest geneticists. Their promising research was brought to an abrupt halt, however, by the rise to party favor of T. D. Lysenko, a biologist who subscribed to Lamarckism (the opposing theory to Darwin's), which held that acquired traits could be inherited. (Giraffes have long necks because of generations of straining for upper branches, and so on.) Not only did this deny Darwin's natural selection as the evolutionary engine, but it was also incompatible with Mendel's theory of genetics.

Rather than take on the Mendelians with scientific argument, Lysenko adopted the more effective strategy of political banishment. He branded his opponents idealists and obstructers of the revolution. Even worse, the Mendelian system they espoused—self-copying genes and random mutations—worked against the Party's plan to control and change nature, in particular, human nature. Once Lysenko had convinced Stalin that Mendelian genetics threatened communist ideology, the science was virtually outlawed in Russia. Russian geneticists who had worked for years with experiments that refined and illuminated Mendel's laws were not merely bullied into silence like American hereditarians during the same period, they were either forced to leave the country or were tried and imprisoned.

Lysenko and his politically correct, but bogus, theories survived Stalin's death and maintained an iron grip on Russian science for *three decades.* He was finally brought down, but only when agricultural dictums he imposed on Russian farmers led to the crop failures of 1964. It is widely believed that Lysenko's disgrace contributed to the fall of Nikita Khrushchev, who supported his theories. With Lysenko gone, Mendelian genetics returned to Russia, and Lysenko's work is now totally discredited by Russian scientists. The sorry tale of his rise and fall stands as a grim warning about the politicizing of science.

AMERICA'S ENVIRONMENTALLY ORIENTED social scientists did not need the mailed fist of a Stalin to triumph over their opponents. With such forceful intellects as Boas, Kroeber, Mead, Benedict, Watson, and Skinner fighting their cause, they sailed to victory, greatly aided by the political winds at their rear. Heady with success, they made ever wilder claims for the environment's molding power over humans—and made ever more disdainful dismissals of genetic contributions. As with many who announce new concepts about human behavior, the innovators, emboldened by the cordial reception their insights received, leap from moderate positions of partial environmental influence to sweeping pronouncements of total governance. And if the founders themselves didn't raise the stakes in this way, their disciples were sure to do it for them.

This dynamic held true for the theories of Freud, Boas, and Watson. In the latter case, Watson did the escalating himself. While Watson's writings continued to acknowledge inherited traits, his euphoria at having unlocked the secrets of human behavior moved him to make the now famous statement: "Give me a dozen healthy infants, well-formed, and my own specified world to bring them up in and I'll guarantee to take any one at random and train him to become any type of specialist I might select—doctor, lawyer, artist, merchant-chief and, yes, even beggar-man and thief, regardless of his talents, penchants, tendencies, abilities, vocation, and race of his ancestors."

To make model humans, Watson, in effect, needed nothing from genetic endowment. All he needed was an individual who was "well formed"; he could do the rest. If this sounds like Henry Higgins run amok, it should be remembered that Watson was no isolated visionary but the leading psychologist of his day, one who had reversed the course of American psychology as laid out by William James along Darwinian lines of inherited instincts. Watson's theories, with an assist from B. F. Skinner and others, became the reigning orthodoxy in America on human behavior and remained so for many years.

The three dominant themes on behavior for a good part of the century were Freudianism, which said aberrant behavior was produced in the childhood environment; Boasism, which said behavior was produced by the cultural environment; and behaviorism, which said behavior resulted from environmental conditioning and learning. All were united in enthroning the environment as the determinant of human behavior and in relegating biological inheritance to insignificance. This three-pronged environmentalism was the accepted wisdom that was taught in all universities and that informed serious writing on human behavior—social problems, psychological problems, mental illness—or normal child development. Professor Higgins may have run amok, but he had also taken over—and remained in control until only recently.

THIRTEEN

SHORT AND HAPPY LIFE
OF THE TABULA RASA

THROUGHOUT THE DECADES of environmentalism's rule, the biological view of human nature survived, but its adherents were either driven into silence or they changed the subject. The few who persisted in airing theories about "instincts"—or any other term for inherited behavior—were all but ignored, or considered too fringe for polite society. Shortly after the end of World War II, this group made what may have been a stab at a comeback. In September 1946 a conference entitled "Genetics and Social Behavior" was organized in Bar Harbor, Maine. In the context of the day's intellectual certainties, that was akin to holding a conference today on the scientific applications of astrology.

Knowing the unpopularity of their views, the participants were perhaps emboldened by the respectability that the biological perspective of behavior still enjoyed in Europe, where behaviorism, like Freudianism, never achieved the supremacy it enjoyed in America. The Watson/ Skinner doctrine had such a tight grip in the U.S. that the Maine conference proved no threat to its monopoly on psychological wisdom. Historically, the meeting's significance is not that it launched an opposition movement but rather that it offered a rare glimpse of the day's handful of social scientists who rejected the dictates of behaviorism and cultural determinism.

The first counterdevelopment arrived during the 1950s, when word came from Europe of intriguing animal-behavior studies, most particularly those of Oxford's Nikolaus Timbergen and German ethologist Konrad Lorenz. All of their research—which emphasized mating patterns, pecking orders, territoriality—offered fascinating information on the complexities of animal behavior but always with obvious parallels to human behavior. At the same time that these new ideas were intriguing European circles, the discovery in 1953 by another Watson, James, and his Oxford colleague Francis Crick of DNA's double-helix structure opened the door to vast new areas of genetic research. It would, however, be another twenty years before the scientific world gave broad consideration to the link between Crick and Watson's twisted-ladder molecule and human behavior. And today, forty-six years later, there are still scientists who deny the connection.

Throughout the 1950s and 1960s, the animal-behavior studies of the European ethologists were quietly, almost surreptitiously, advancing the notion of biological underpinnings to *human* actions and social arrangements that had long been thought cultural. Although the authors avoided talk of humans and stuck to safe, neutral species, the relevance to humans jumped from every page and launched the slow, massive shift in how we thought about ourselves. It should be pointed out that the ethologists owe a portion of their success to an interest in their work from the enemy camp, prominent members of the environmental and cultural-determinist school. Some leaders of this school, such as anthropologist Alfred Kroeber, were rethinking their strict environmentalism and, in an admirably ecumenical sense, reopened the door to alternative explanations of human behavior that they had been instrumental in closing.

One reason for their open-mindedness was the setbacks of the behaviorist school. Among the group's "proofs" that behavior was learned rather than innate were experiments that had altered animal behavior by a series of rewards and punishments. The mere act of changing an animal's behavior was taken as hard evidence there was nothing innate in it. To the dismay of the Watson/Skinner group, it was found that in the months following the experiments the reeducated animals invariably returned to the behavior they had been conditioned to abandon. The wholesale backsliding on the part of stubborn rats and mice had upsetting implications for the no-instinct, blank-slate crowd, who could not easily explain the animals' recidivism as nostalgia for tradition. Although some behaviorists were willing to grant instinct to animals but denied instinct in the human, Watson and his followers had claimed that for animals as well as humans, instinct was of no importance; learning was everything. But if that were true, what made the animals revert to their old patterns?

One of the strongest blows to beleaguered behaviorism came in experiments with rhesus monkeys conducted by Harry Harlow of the University of Wisconsin that were reported to the American Psychological Association in 1958. Harlow's experiments involved infant monkeys that were offered two alternative "mothers," both of which Harlow had constructed. One mother was a wire contraption that dispensed milk; the other was a cuddly monkey made of terry cloth that had no milk, only softness. Despite the tasty reward from the wire mother and the lack of one from the cuddly mother, the baby monkeys all clung to the terry-cloth mother.

The infant monkeys' choice was bad news indeed for the behaviorists in that it demonstrated clearly that punishment and reward were not the whole story, as Skinner and Watson claimed for both animals and humans. Something else was clearly going on within the monkeys, in this case it appeared, an innate preference for warm and cuddly. It also carried a relatively new message that the "something" built into the baby monkeys was not just a sweeping trait like ranking-order or territoriality but a trait as specific as a predilection for a particular physical form—which, incidentally, few humans could deny having in another context. Once again monkeys had made monkeys of the behavioral theorists. (Harlow's discovery might also suggest something about human children's affinity for teddy bears and other furry toys.)

Like most American psychologists who experimented with animals, Harlow worked with monkeys that had been reared in captivity. The European ethologists, on the other hand, studied animals in their natural habitat. This would become an important difference. Eighty years after Darwin, humans were still having difficulty with the concept that they were just another animal, that they shared behavioral patterns with rats and mice. The monkeys' artificial existence in laboratories was seized on by the resisters as a face-saving explanation. These lab creatures, after all, were reared in a totally artificial environment and may have been made more human by prolonged human contact. Differences from their free-roaming cousins had, in fact, been noted. We could shrug off any relevance to human behavior of these laboratory hybrids, who were clearly the Uncle Toms of their species, faking human traits and otherwise playing along to please their human masters.

The excuse was short-lived. When animals in the wild were also found to behave and organize themselves in complex ways—regardless of contact with each other and with all but no contact with humans— a conclusion of built-in behavior mechanisms became unavoidable. Lorenz and Timbergen worked with creatures like bees and geese that were relatively easy to observe in unmolested states, but their research led to more ambitious ethological studies, such as Jane Goodall's years in the wild with chimpanzees and Dian Fossey's with gorillas. All of the fieldwork left no doubt about the instinctive nature of animal behavior and social organization, some of it highly complex—and strikingly, tellingly, poignantly close to our own.

IN 1961 a curious and remarkable book burst upon the American consciousness. It was titled *African Genesis,* and it was written by a playwright and screenwriter named Robert Ardrey, who had become fascinated by the paleontological research then in progress on man's African origins. Abandoning Broadway and Hollywood in 1956, Ardrey went to Africa and set out to educate himself in this science. The book that resulted five years later provided an overview of all the animal-behavior research that had been progressing quietly for the preceding twenty years, particularly that which focused on territoriality and social hierarchies. Ardrey emphasized that within species after species a powerful aggressiveness was an instinctive part of a territory-status-sex "bundle of instincts." The ethological and zoological groundwork led him to his main thesis: Humans are a product of this possessive-combative lineage, that we are descended from killer apes from whom we have inherited strongly aggressive, even homicidal inclinations.

African Genesis was written in a highly colorful, exuberant, often lyrical style and was packed with intriguing evidence—tasty gossip about our fellow animals—to support Ardrey's thesis. A vast audience was gripped to read about the scrappy, status-obsessed natures of mead-owlarks, field mice, and tropical fish. It turned out, for instance, that the warblers' song was not a lyrical burst of joy at the beauty of nature but a surly warning to other warblers to stay out of the singer's territory.

When Ardrey arrived at humans, he cited at one end the fossil evidence of early human nastiness that left no doubt about within-species foul play, that is to say, homicide. At the other end, he pointed to mankind's sorry career of bloodshed. Speaking of the earliest humans, he produced scene-of-the-crime evidence that lethal weapons go back to the very origin of the species. One of the most vivid was the skull of a young male *Australopithecus* that had clearly been cracked by a blunt instrument. On the basis of such clues Ardrey proclaimed that the human race had been born with a weapon in its hand and has been wielding it with gusto ever since. Hello, N.R.A.

Ardrey's powerful and dramatic assertion of inherited behavioral traits boldly confronted the behaviorists, still solidly in power. And not only was this screenwriter challenging their dictums that behavior was learned, he was also saying that our inherited behaviors were nasty. Given the utopian, all-babies-are-good cast to the behaviorist canon, this news was a doubly unpleasant jolt. It was bad enough to hear that we might be born with *any* behavioral leanings; but that those leanings included homicidal violence was not a happy message for those who saw our species as a higher form of life bursting with morality. It also did damage to hopes for a peaceful, more civilized world.

In spite of the disagreeable message, Ardrey's thesis, for many, struck a plangent chord of recognition and seemed to hold out a more realistic, more insightful explanation of man's endless warfare than historians who pondered what had gone awry at the Congress of Vienna or the Versailles conference. *African Genesis* was widely read, and its ideas about our killer-ape heritage seeped into the public consciousness. Even more significantly, it marked the beginning of wide-scale consideration of inherited aspects of human behavior. It was a Fort Sumter shot fired at the tabula rasa bulwark behind which the social sciences then operated. This intellectual shift has gained momentum ever since.

While a highly successful screenwriter and a respected dramatist, Ardrey had no scientific credentials whatsoever. Although his amateur status may have liberated him from academic caution and emboldened him to state forthrightly the implications of others' research, it permitted his critics, often people who *did* have scientific credentials, to treat him with disdain. He ran smack into the problem faced by all who promulgate new ideas; their writing more often than not is appraised by experts in the field, people deeply wedded to the very ideas the new work is challenging. The experts the rest of us might assume would be the most stimulated and inspired by new ideas relating to their fields are far more likely to marshal their expertise to fight off the upstarts. A sorry rule is that fresh theories are rarely weighed on their merits but instead trigger displays of intellectual territoriality that guarantee a fiery gauntlet for anyone threatening prevailing orthodoxy. Since most men and women engaged in the idea arena know this nasty fact, few are willing to jeopardize respected professional niches with disruptive theories. It almost requires an amateur like Ardrey—in his case, one with a talent for vivid, gripping communication—to undergo the inevitable pillorying. And pilloried he was. Knowing how influential his book had been, I was astounded to go back and read the reviews, almost all of which were by experts in the field and almost all of which were scathing.

If Ardrey was subjected to the contempt of scientists, he was no wimp at returning it. In a later book, he presented a summary condemnation of the environmentalist dogma that is a fine example of killer-ape vitriol. His academic opponents present for him a picture, he says, of "cultural anthropology, behaviorist psychology, and environmental sociology like three drunken friends leaning against a lamppost in the enchantment of euphoria, all convinced they are holding up the eternal light when in truth they hold up nothing but each other."

Ardrey's thesis about humans' built-in aggressiveness was anathema to the many who had been brought up to believe Margaret Mead and her colleagues' conclusion: We could have been as sweet and lovable as the Samoans of Mead's imagination if our corrupt societies, our rotten cultures, hadn't turned us mean and vicious. To preserve this vision, we today have the more specific whipping boy of television and film violence to blame for our lost Samoa. At the time of *African Genesis*, however, murder rates were down and our country hadn't been slaughtering foreigners in any major way for a peaceful sixteen years, so to the middle-class academics who had never killed anything more than the random roach or mouse the suggestion we might have innate homicidal leanings was an absurd slander.

So repugnant was Ardrey's charge, and so counter to the prevailing orthodoxy, that not only reviewers but also paleontologists and other academics rose up in angry phalanxes to fight him with a territorial desperation that would have impressed Siamese fighting fish. One point they seized on had to do with human teeth. Important to Ardrey's man-the-killer argument was that our species started out as meat eaters, a fact he felt was proven by the visible wear on fossil teeth. But this proof, in the eyes of many experts, was anything but conclusive; they produced counterevidence to show that we had started our species career as vegetarians—not that this diet would mean that we were nice, but probably nicer than a species with an urge to rip apart animals with its teeth.

In *The Third Chimpanzee,* his fascinating 1992 book on evolutionary psychology, anthropologist Jared Diamond sums up the counterarguments to Ardrey in an early essay that cites the shaky paleontological grounds for believing *Homo sapiens* had always been carnivorous, that our fondness for lamb, beef, and veal chops might not be something we picked up from Julia Child but has been with us from the get-go. As Diamond puts forth the rebuttals, he picks up the what-does-he-know-about-science tone typical of Ardrey's critics. Later in the same book, in an essay written some years later, Diamond examines mankind's sorry history of homicide, presenting a plethora of grotesque statistics that would seem to place killing high above altruism and incest taboos for a cross-cultural, specieswide trait. With admirable candor, Diamond acknowledges that Ardrey appears to have been right in his main thesis; we have a homicidal genetic heritage. Wrong about the teeth, he in effect says, but right about our nasty nature.

Whether or not we started out as meat eaters, the evidence is everywhere that we sure do like to kill. The frenzy with which the opposing scientists struggled against Ardrey's construct was almost touching. They seemed to believe that a different interpretation of the grinding marks on the *Australopithecus* incisors would expunge two thousand years of planetwide bloodshed and render us a gentle, agreeable species that had been turned bad by the Industrial Revolution, Swiss bankers, the Pentagon, Madison Avenue, *The X Files*, lead poisoning, and a host of the usual environmental suspects.

In spite of the critical trouncing of *African Genesis*, it still received major attention from the most influential and widely circulated publications. Not only was the book reviewed everywhere, but also the critiques, although mostly negative, often were the lead reviews, an honor usually reserved for books the editors consider important. This conflict between serious attention and negative assessment suggests that while the scientific reviewers had no use for the book, the nonscientist editors were fascinated.

For all *African Genesis*'s groundbreaking novelty, Ardrey in his first chapter acknowledged his debt to the animal-behavior research of others. He cited especially W. C. Allee, a zoologist at the University of Chicago, whose 1940 book *The Social Life of Animals* posited the universality of a dominance instinct. Ardrey credited another book with initiating the entire train of thought about inherited behavior: a 1920 work by an English bird-watcher, Eliot Howard, entitled *Territory in Bird Life*. He also gave credit to an American zoologist, R. C. Carpenter, who "brought matters perilously close to home" with his studies of territoriality in apes and monkey societies living in natural states.

While this research may have made a profound impression on Ardrey, few others paid much attention. Similarly, the work of Lorenz and Timbergen, which he also cited, was of interest only to a small group of psychologists. Because this work dealt exclusively with animals, the psychologists could take from it whatever implications for humans they wished. Konrad Lorenz, in his 1952 book *King Solomon's Ring*, draws a few mild parallels between animal and human behavior, but he doesn't hammer the point as Ardrey does. For the most part, however, until Ardrey, the environmentalists had convinced the world of the irrelevance of instinct to humans, and the animal behaviorists were too timid to insist on it.

By contrast, *African Genesis* forced widespread attention to the ethological premise—that we can learn much about ourselves by studying animals. The Boas/Watson/Skinner forces, with their vision of a human unencumbered with behavioral predispositions, had succeeded in disconnecting our species from the rest of evolution; almost single-handedly, Ardrey linked us back up. Not only was Ardrey, with his three-million-year-old unsolved murders, claiming that evolution has saddled us with a battery of behavioral traits, but he was also reckless enough to emphasize the most repugnant, the killer impulse. This inflammatory claim certainly won Ardrey attention, but the angry controversy it provoked almost obscured his main point: that human behavior is as much a product of evolution as the human body. This generalization was so much more palatable than his specifics about killer instincts that it almost slipped by with little fuss. A great tectonic plate of intellectual thought had lurched into motion and has been picking up speed ever since.

In the next few years a number of ethological books emerged that had comparable, even greater impact: Lorenz's *On Aggression* (1966), Desmond Morris's *The Naked Ape* (1967), Lionel Tiger's *Men in Groups* (1969), and *The Imperial Animal* (1971), which Tiger wrote with Robin Fox. All of them emphasized the major role played by genetic inheritance in human behavior. Lorenz's book was treated far more respectfully than Ardrey's, undoubtedly because of his eminence as a scientist (he won the Nobel Prize in 1973), but basically it made the same point about humans' homicidal aggressiveness. All of these books contributed mightily to an evolving awareness among the thinking public about the behavioral legacy that has come down to *all* humans through the eons. Little was said by these writers about *how* it came to us, but everyone knew there was only one way: genetic transmission.

In my own conversion, *African Genesis* was the first of these books that I read, and it made an enormous impact on me. I felt that at last someone was not only talking sense about human nature but also was speculating intelligently on the evolutionary origins of that nature. That Ardrey was my first exposure to this thinking, I assumed, was happenstance, the luck of the library draw; I made no claim to be on top of every new development in ethology and anthropology. I assumed *African Genesis* appeared about the same time or after the other books. When, however, thirty years later I began researching the subject in earnest, I was stunned to learn that Ardrey's book, which made as forceful a statement about our inherited behavior as any of them—and was far more widely read—preceded the others by *five years*. Hooray for amateurs!

While *African Genesis* had a bumpy road to its eventual impact, Ardrey's next book, *Territorial Imperative*, was a solid hit from the start

and enjoyed a long run on the *New York Times* best-seller list. It bene-
fited not only from the millions of readers who had been excited by
African Genesis but also by the near simultaneous publication of Konrad Lorenz's *On Aggression*, a highly readable yet scholarly book that
provided strong validation for Ardrey's ideas from a world-renowned
scientist.

My awe at Ardrey's audacious accomplishment continued to grow.
When I interviewed Lionel Tiger for this book, I asked what drew him
to the subject of evolved behavior in humans. Because he was a
respected scholar, I assumed he had been researching the subject for
years before the ethological books, his and the others, burst onto the
scene in the 1960s. A professor of anthropology at Rutgers and the
author of many papers and books, Tiger had all the scientific credentials that Ardrey lacked; I had believed him to be a pioneer in placing
evolutionary-behavioral thinking before the public. I was therefore
startled when Tiger told me his interest had been stirred by his reading
African Genesis, that the book had been a major factor in his shift to
ethology and the evolutionary perspective on human behavior.

With Ardrey way out in front, a civilian Joan of Arc leading a handful of seasoned generals, the combined influence of all these books
was powerful indeed on the reading public. Together they shattered
the fortress of behaviorism and open the gates to consideration of our
species' innate behavioral characteristics. The willing, almost eager
acceptance of this concept was reflected by the appearance in the public discourse of terms like *territoriality*, *pecking order*, and *fight-or-flee*
impulses.

I suspect that the quick adoption of these terms into the vernacular
said more about the broad recognition of their relevance to humans
than to their novelty or wit. Has anyone worked in an office and not
perceived a pecking order, or heard loud music from the next apartment without feeling a murderous surge of territoriality? I further
believe that throughout the environmentalist heyday, the public at
large paid little attention to the dictums of top psychologists and knew
full well that humans were born with personality traits and behavioral
dispositions. The ethological books of the early 1960s only provided
scientific validation to long-standing popular wisdom. Whatever resonance the books had with the general public, they initiated a gradual

migration of many thoughtful people to a balanced view of a human nature, a nature shaped by *both* the environment and genes.

A favorite jab of the critics is that the biological perspective reflects conservative political thinking and resurfaces during zeitgeist swings to the right. It is therefore worth noting that the resurgence of behavioral genetics thought occurred during the 1960s, one of the most politically liberal periods in the country's history. The pervasive mood of rebellion did nothing to slow the advance, nor did the day's progressive spirit provide any tailwinds for the new view of human nature, the kind of assist environmentalists received from the progressive movement earlier in the century. Ardrey, Lorenz, and the others made such convincing cases for the human inheritance of behavioral patterns that more and more people acknowledged the point's validity without brooding about the political implications or jumping to conclusions about dire ramifications for society. To many, the ethologists were offering plausible explanations for behavioral phenomena that were all too apparent.

None of the books claimed that inheritance was all-powerful in human behavior, only that we had certain vestigial impulses that operated in concert with environmental influences. The sensible balance of this view, however, was quickly threatened. As with so many new ideas that finally take hold, the genetic perspective appeared to be careening toward overkill. After only a few years of acceptance, the new thinking seemed to be lurching toward an exclusively biological view of behavior. Because of the forty-year moratorium on such thinking among educated people, the inherited side of human nature was the exciting news of the day, and the news was being seized upon as a thrilling new perception that would *replace* environmentalism. Non-scientific kibitzers wanted a fight with a winner and a loser. This led to many mischaracterizations and overstatements in press coverage. Responsible scientists held firmly to the concept of a draw, of innate nature *interacting* with nurture.

Even this modest adjustment to basic assumptions about behavior, many knew, necessitated discarding, or at least rethinking, most of the psychological research done in the preceding decades. In spite of the wasted effort and misspent careers, many took heart in the belief that the study of the human was at last on a valid track after forty years of radical environmentalism. The new push to disentangle genetic from environmental influences was called behavioral genetics. To none joining the new quest did this mean the environment no longer mattered but only the door had at last been opened to one half of the human equation that had been ignored for fifty years.

FOURTEEN

SURVIVING THE JENSEN FUROR

THROUGHOUT THE 1960S, the new train of thought—gene-influenced behavior—was lurching forward, picking up a number of important passengers, and signaling to believers that the behaviorist tyranny was finished and that they could emerge from hiding. The hard-won receptiveness to genes was, however, abruptly derailed in 1969 by the publication of Arthur Jensen's paper that claimed I.Q. differences between blacks and whites were not explained by cultural differences but were, in fact, genetic. The resulting explosion all but killed off the infant field of behavioral genetics and certainly set it back ten years. Since the same debate erupted again in 1993 with the publication of *The Bell Curve*, it might be useful to review the earlier conflagration.

Throughout the decades when environmentalism ruled, the incendiary subject of I.Q. managed to slip by the moratorium on talk of inherited traits. It was granted a special status, separate and distinct from other aspects of behavior, or at least on the border between a personality trait and the physical self. Perhaps one reason for the dispensation was that intelligence tests had grown increasingly refined and their validity was generally accepted by educators who found them useful in assessing whether or not little Jane or Johnny should be admitted, or once admitted, if they were doing their best. Whether or not this thing that was measured, called I.Q., or "g," was inherited, was not central to the educators' pragmatic need for some sort of classifying test. As a result, the controversial point was not pushed. I.Q. malleability or lack of it was certainly relevant to educational strategies, but few psychologists felt the data was conclusive enough to insist on a genetic basis.

The tests, however, were never accepted by the radical environmentalists, who saw concessions about basic cognitive abilities as validation of the hereditarian viewpoint. To them, affixing students with allegedly innate I.Q. numbers was an unfair and indelible stigma. From such categorizations, it would be a small step to assigning everyone to an economic niche within society from which they had little hope of escaping. I.Q. was seen as another device for the haves to keep down the have-nots.

The environmentalists insisted that intelligence was a product of rearing circumstances and could be improved. They had to reach far for examples of environments that had actually raised I.Q., and while they found some, it was far outweighed by evidence of an innate mental endowment that varied greatly between individuals but varied little within the lifetime of most people. The extreme opponents of I.Q. denied there was such a thing as intelligence, insisting the entire concept was an invention of reactionary scientists to place social injustice within the natural order.

Arthur Jensen was a highly respected psychologist at the University of California who had worked for years in the area of I.Q. Because of his broad knowledge of group performance on tests, he felt obliged to speak out about recent governmental programs like Head Start that were aimed at improving the scholastic performance of underprivileged children, most of whom were black. Jensen's paper on the subject appeared in the winter 1969 issue of the *Harvard Educational Review* and took the position that such programs were misguided in that statistical evidence demonstrated consistently lower I.Q. among American blacks than among whites; Jensen went on to try to prove that these differences were not caused environmentally but genetically. It was therefore futile, in his view, to expect government programs to alter this fact of nature. Jensen's essay turned out to be the little paper that launched a civil war.

For years psychologists had known that black children in America had measured on average fifteen I.Q. points lower than white children. No one argued this statistic. The fact argued passionately by everyone was *why*. For years the generally accepted answer, certainly the most palatable to fair-minded people, was that deprived environments caused the discrepancy. Now one of the most respected psychologists in the country, a specialist in I.Q., was offering hard evidence—test scores, factoring out cultural bias, comparisons with other culturally deprived groups, and so on—that the cause did not lie with the children's environment but with their genes. The grim corollary was that there wasn't much to be done about it.

The paper outraged just about everyone. Even people sympathetic to the advances of behavioral genetics and the growing interest in the biology of personality now saw ugly racist implications to this pursuit, and they withdrew their support of behavioral genetics in general, some out of a repugnance at racism, others out of fear of association with a stigmatized science. Eminent scientists in other fields who had shown little interest in the developments in ethology and behavioral genetics now came down hard on those who had. It was one thing to talk about aggression in field mice, even in humans, but quite another to talk about racial inferiorities. Some who were willing (behind closed doors) to allow a degree of validity to Jensen's analysis were angered by what they saw as a hurtful and irrelevant digression.

The environmentalists, who had warned of the dire implications of behavioral genetics, were delighted by the furor and led the pack in denouncing Jensen with a we-told-you-so relish. Undoubtedly dismayed by the growing acceptance of genetic thinking, they saw in Jensen's paper an opportunity to rip the mask of respectability from the behavioral geneticists and expose them as racist reactionaries. Fusillades of papers and articles appeared, the authors each striving for greater outrage than preceding writers. Even pioneer behavioral geneticists like Gerald Hirsch at the University of Illinois brought all their science to bear in attacking Jensen's analysis of the data.

The attacks focused less on Jensen's science than the motives behind it. Those who believed in a genetic contribution to personality were not scientists of good will who had strayed into error; they were political operatives of the far right who had concocted bogus data to further a reactionary agenda. It would not be overstating it to say that the backlash against genetic thinking reached almost hysterical proportions. At a time when thrilling advances were being made in civil rights, it was for many a nightmare to witness a reputable scientist appear to say blacks were a lost cause. Eugenics, for all the alarm it caused in its critics, was granted a thirty-year free ride compared to the instantaneous uproar caused by the Jensen paper.

In the early 1970s, Robert Plomin, who was then new to the field of behavioral genetics, attended his first major professional conference, a meeting of the Eastern Psychological Association in Boston. The keynote address was given by Leon Kamin, already a distinguished psychologist and Marxist who, according to Plomin, was "leading the charge against behavioral genetics." Plomin described the scene: "Before an audience of thousands, Kamin thundered at length about the evils of behavioral genetics and brought the crowd to its feet. The address . . . was an *ad hominen* attack on behavioral geneticists and their political motives. I was shocked by what seemed emotional rabble-rousing and stunned by the welcome it received."

Eventually this sort of furor subsided, but only when the hereditarians beat a retreat, leaving the environmentalists once again in control of, if not behavior in general, at least intelligence. With calm restored, few of the combatants doubted that Jensen had done what he thought was his responsibility as a scientist. Everyone who knew him, including most adversaries, agreed that he was a solid scientist and a decent, fair-minded individual with no ill will toward Afro-Americans. When, however, he saw public policy being based on assumptions that his years of research had told him were false, he felt he had an obligation to speak. Still, in looking back on the resulting mayhem, it is hard not to see his action as precipitous and destructive.

Most people in behavioral genetics know what a minefield lurks in the entire subject of group differences. Not just racial differences, but gender differences and ethnic differences—even age differences—can infuriate many. Most behavioral geneticists stress that group differences are a subject quite separate and distinct from their area of interest, which is *individual* differences. They frequently point out that within any group there is more variation among individual members of that group than there is between any two groups. Even Jensen had pointed out the average I.Q. variation within raised-together siblings is as great as the average difference between whites and blacks. This makes thinking along group lines when confronting an individual highly risky; in fact, it makes it altogether irrelevant to a fair assessment. A clear understanding of behavioral genetics, they say, should make a person less racist or sexist rather than more.

In addition, as a human designator, race is an imprecise, unscientific concept—as it increasingly appears that gender is as well, if to a lesser degree. Not only are Afro-Americans invariably mixtures of races; they are also mixtures of different strains of Afro-Americans. And race itself, it is known, results from a complex combination of genes and not always the same complex within such vague race designators as Afro-American. Also, as evidence mounts for the degree to which performance can be affected by self-esteem, lower I.Q. scores can be self-perpetuating statistics, especially in disdained groups.

For all the explosiveness of the subject of race differences, and for all the importance any differences might have for social policy, it is, scientifically speaking, a relatively small point in the context of the vast subject of human personality and behavior. For a field that promises to reveal causes of depression, alcoholism, violence, homosexuality, mental retardation—in addition to one day answering such venerable questions as the extent to which environments can alter behaviors—for this exciting, hope-offering field to become mired in excoriating arguments over whether group A is smarter than group B can only be termed a tragedy for our culture's intellectual life.

For this reason I feel strongly that aside from whatever scientific weaknesses existed in Jensen's analysis, the subject should have been avoided because of the damage it did to a new field of valid, highly promising research. In the late 1960s, behavioral genetics was an infant science. Years of domination by an opposing theory had stifled it; but thanks to research in such related fields as ethology, as well as the accumulation of failures in behaviorism, behavioral genetics was beginning to move into the intellectual mainstream. Almost immediately an eminent scientist made sweeping claims, claims that were extremely wounding to a portion of the population, about a trait that was not fully understood, perhaps of marginal importance, and any mention of which was inevitably explosive.

FOR ALL THE DAMAGE inflicted by Jensen's paper, behavioral genetics projects quietly got under way at the universities of Colorado,

Minnesota, and Louisville, as well as at a number of smaller institutions. Such major research initiatives, however, remained all but out of sight and were happy to be little noticed. In most conduits of ideas— the media, college courses, new books—there were few allusions to a genetic contribution to personality and behavior. Serious discussions of behavioral traits reverted to the purely environmental theories of the 1950s and 1960s. Once again the subject of genetic influence was driven underground; the stench of racism was too strong.

Gradually, the furor subsided. The environmentalists appeared to have recaptured their invaded realms, an impression that owed more to the embarrassed silence of geneticists than to a decisive intellectual victory. Not everyone was silent, however. In the July 1972 issue of the *American Pyschologist* a curious document appeared. It was a manifesto of sorts, signed by fifty top psychologists and other scientists, including four Nobel laureates, deploring the "suppression, censure, punishment, and defamation . . . against scientists who emphasize the role of heredity in human behavior." It proclaimed the importance of the new information about this role and the determination of the undersigned that research into the hereditary influences on behavior should proceed. The defiant cri de coeur, however, had only slight impact and did little to disrupt the regained hegemony of the environmentalists. (A historic curiosity, the letter is reprinted in appendix A.)

A far bigger threat to the environmentalists' peace of mind was the publication in 1975 of Edward O. Wilson's landmark book *Sociobiology: A New Synthesis*, which reopened the genes versus environment debate at a level of acrimony that suggested the wounds inflicted by Jensen were far from healed. Wilson, an esteemed professor of zoology at Harvard, had produced a six-hundred-page summary of the wealth of ethological research that had been accumulating over the past twenty years, particularly the influential ideas of W. D. Hamilton on kin selection and R. L. Trivers on parental investment—theories that went far toward answering questions about natural selection that had been left unanswered by Darwin. The book was densely technical, with pages of charts and statistics documenting the elaborate social behavior of various species.

Few social scientists had difficulty accepting the 95 percent of the book that dealt with animal behavior, and the meticulous presentation of research data would have made attacks difficult. In the last chapter, however, Wilson reviewed existing knowledge about early *Homo sapiens*, but then spoke of the mission of sociobiology "to reconstruct the history of the machinery" (by which he meant the evolution of the human brain) and "to identify the adaptive significance of each of its functions." Sociobiology, he wrote, must also "monitor the genetic basis of social behavior." He did not feel it necessary for future sociobiologists to prove there *was* a genetic basis to human behavior, simply to monitor it. What to Wilson appeared self-evident, or rendered unavoidable by his book's previous 574 pages, was still hotly contested by the environmentalists, as he would quickly discover.

The outrage was immediate and vehement. Harvard's Richard Lewontin and Stephen Jay Gould were among fifteen prominent academics who signed a letter in the *New York Review of Books* that denounced *Sociobiology* and drew parallels with racism and Nazism. Wilson's lectures were picketed, and students heckled him as he tried to speak. At a scientific symposium, he was physically attacked. At the 1976 meeting of the American Anthropological Association, a motion nearly carried to censure Wilson's book. (Among those present was Margaret Mead, who also evoked Nazism, but in Wilson's defense, when she accused her fellow anthropologists of "book-burning.") Looking back at this tempest, it should be remembered that Wilson's crime was nothing more than suggesting that aspects of human social and cultural institutions may have evolved through genes. For that heresy he was not to be disproved or disagreed with, but *censured*. Galileo must have chuckled.

To the environmentalists, rats, mice, and other animals were okay as long as they could learn tricks, but once they evinced stubborn, instinctive behavior they became dumb animals, irrelevant to the human condition. The animals of Wilson's studies were riddled with instincts that propelled them into an array of complex actions and social arrangements disturbingly close to those of humans. Once again the antihereditarians hit the alarm button. The group had remained relatively quiet when the ex-screenwriter Ardrey had proclaimed man's behavioral links to feisty chimps and apes; they were even able to tolerate Lorenz and Timbergen as Europeans who had perhaps spent too much time among their geese and honeybees. As for the brilliant and well-received *Imperial Animal*, how seriously would any-one take a book on animal behavior written by professors Tiger and Fox? (Many did.)

Wilson was a different matter. He was a respected naturalist who spoke from Harvard's zoology department. His credentials were unas-sailable, and he had laid out in exhaustive detail the research upon which he drew. His conjectural conclusions about humans, however, provided the critics with the Achilles' heel they needed, and they attacked it as if the book's other twenty-six chapters didn't exist. One hundred and four years after Darwin's *The Descent of Man*, leaders in the social sciences were still angrily rejecting the proposition that human behavior, like that of other species, was influenced by a genetic heritage. Now, Alfred Wallace must have chuckled.

Despite the attacks, the new field of sociobiology took hold and began turning up a series of fresh insights about human behavior. Psychologists who subscribed to its genes-based view could be seen as returning to concepts that had once flourished in their discipline but had lain dormant for years. A new development was that scholars from other fields, such as economics and political science, were for the first time contemplating their disciplines from an evolutionary perspective.

By the late 1970s, the academic air was braced by a sense of a new and penetrating paradigm, one that promised a more reality-based conception of human behavior than had been possible with earlier all-or-nothing models. Because of the spreading excitement over the "new synthesis" that Wilson had forged, it is odd that his term *sociobiology* fell into disfavor. In the eyes of well-wishers, this was a result of the excesses of early sociobiologists, who, stimulated by their newfound truth, set to explaining everything from car wrecks to stock-market dips in terms of Pleistocene exigencies. The term *sociobiology* fell from use—but only the term, not the discipline. Like felons who have served the time for early mistakes, the field changed its name before going straight. It continues to flourish under a number of terms, but the one that seems to be winning out—there are fashions in these things—is *evolutionary psychology*.

Whatever they called the line of thought initiated by the etholo-gists and Wilson, it changed forever the concept of culture as some-thing man-made, or as Boas believed sui generis. It was now seen increasingly as having originated in our genes. Although this simplistic formulation was immediately made complex by Wilson himself, who said in 1983 (in a paper written with Charles Lumsden), that "culture is created and shaped by biological processes while the biological processes are altered in response to cultural change." They were exchanging the old circularity (culture creates culture) for a more plausible model: Culture, which originates in the human genome, *edits* culture.

I welcomed this analysis of culture's origins. If we can quickly allow that animal "cultures" are genetic, why was it so out of the question to think ours, at least in part, might be similarly rooted? I had long been bothered by the assumption of human cultures as entities that had always been there and whose origins were out of reach. When, for example, people discussed our tradition of monogamy as "cultural," what exactly were they saying? Did they envision a committee of gray-bearded rabbis somewhere in the Sinai around 3000 B.C. saying, "How about we allow each man one wife? Does that fly with all of you?" It's one thing to shrink from such imponderables as the universe's first bang, or the origins of life, but to duck the origins of culture—with all we now know about the complicated living arrangements of other species—is to me intellectually flaccid and serves only to avoid the most plausible conclusion: Human cultures have been shaped and altered by historical contingencies but originally grew out of genetic dispositions. Like penguins but unlike walruses, we favor one mate.

BY THE LATE 1970s the old anthropological view of purely man-made cultures was gone forever. It took somewhat longer to kill off the other great twentieth-century flowering of environmentalism: Freudianism. This monument of modern thought did not exit gracefully but required an array of forces to dislodge its truths so embedded in the century's wisdom. One of these forces was the discovery in the 1950s of mood-altering drugs that opened people's eyes to the chemical basis of mental problems long considered the result of childhood disruptions and the domain of talk therapists. The spectacle of a tiny pill wrenching a person from chronic depression into rosy good cheer was not lost on people who were, at the same time, learning about the nonstop chemical factory within the human body. Reinforcing a chemical view of behavior were the repeated failures of lengthy and costly psychotherapies to cure lingering conditions like depression. Therapists' appeals to reason were getting nowhere, while pills were performing miracles.

The success of drugs was by no means limited to minor problems but was working wonders with the most wrenching mental illnesses, such as schizophrenia and autism. Mounting evidence indicated such severe mental conditions resulted from genetically caused chemical imbalances that pills could correct. Since the Freudians had long claimed that these illnesses were environmentally induced, they had some explaining to do. Instead, they quietly withdrew their environmental explanations—such as the refrigerator mother, castration fears, and other Freudian standbys—and quietly yielded these illnesses to the biological side of the fence. Each year more and more mental and behavioral problems joined the migration out of psychoanalysis and into biochemistry.

The embattled Freudians were further debilitated by squabbles within their own ranks. The endless disagreements among psychoanalysts about even such fundamental matters as diagnostic categories, the frequent public courtroom battles of psychoanalytic experts on opposing sides offering prolix analyses that "proved" contradictory points, ever more strained and improbable nursery scenarios for one ailment or another—all provided outsiders with an impression of a science in disarray. The analysts' lack of success with cures didn't help much either.

Books denouncing psychoanalysis appeared. Standout among them was Jeffrey Masson's account of his training as an analyst; his success in the field, followed by angry disillusionment, left readers with an impression of a profession teeming with, among lesser failings, a charlatanry that frequently reached wholesale fraudulence. Also typical of the negative testimonies, Martin Duberman's *Cures* was a wrenching account of fruitless decades with Park Avenue psychoanalysts who, instead of curing his homosexuality, made him feel worse about it, instilling painful negative feelings for which he paid handsomely with an assistant professor's salary. People like me, who had no such first-hand grievances against the profession, increasingly came to wonder what fifty years of Freudianism had given the world but Woody Allen.

The coup de grâce came in a lengthy piece in the *New York Review of Books* by Frederick Crews in which he systematically demolished not only claims of therapeutic success but claims of psychoanalysis being a science at all. The article was seen by many as an obituary for Freudianism's dazzling career. With the death of this giant the neurological and genetic models of mental illness had all but recaptured the psychiatric profession. Talk therapy was not dead, however. Many people still claimed to have been helped in their emotional problems by various forms of counseling (44 percent of those polled in a 1995 *Consumer Reports* survey), but pharmacology had clearly taken over the therapeutic action. After two thousand years of evil spirits, demons, witches, devils, Oedipus complexes, and castration fears, the world has circled back to the ancient Greeks' view of mental illness's cause: Something was physically wrong with the brain.

BY THE TIME Thomas Bouchard began his study of separated identical twins in 1979, the intellectual climate had decidedly shifted toward an evolutionary, biological perspective on all aspects of human behavior, normal as well as abnormal. In universities across the country, courses specifically on behavioral genetics and evolutionary psychology were appearing on curricula, and few courses of any sort, if they touched on human behavior, now ignored the possibility of a genetic influence.

Faced with this avalanche of setbacks, the radical environmentalists were not ready to admit defeat and clung to the hope that the emergence of biological theories was cyclical. The mounting acceptance of behavioral genetics and evolutionary psychology was written off as an unfortunate fad that owed its reappearance to the conservative zeitgeist, nothing to solid science, and would pass when the political mood changed. It looked like events would bring on this backlash sooner than they hoped.

In 1994 a new call to arms went up to the opponents of genetic applications to human behavior. It was the publication of a book that revived the furor created a quarter century earlier by Arthur Jensen. The book was *The Bell Curve: Intelligence and Class Structure in American Life* by Charles Murray and Richard Herrnstein, which in eight hundred data-packed pages sought to establish (1) the importance of intelligence to success in contemporary life, (2) the genetic basis of intelligence, (3) racial differences in mean I.Q., and (4) the dire implications for our society of these facts and our failure to address them.

Not surprisingly, the book provoked an explosion, in many ways a replay of the 1969 Jensen debacle but perhaps with a shade less vitriol (fewer accusations of Nazism, for instance). As the usual defenders of environmental determinism sprang into action, they seemed to do so with less vigor than they had twenty-five years earlier. Most interesting, however, was the enormous attention given *The Bell Curve* by the mainstream media. The *New York Times* ran a cover story in its Sunday magazine section, then the following Sunday gave the book a lead review in its book review section. *Newsweek* also gave the book a cover story and *The New Republic* ran an article by the two authors outlining the book's main points. In the same issue, however, the magazine took the unusual step of prefacing the article with no less than nineteen brief essays deploring and attacking Murray and Herrnstein's positions, a negative chorus that in the words of the *Wall Street Journal*'s David Brooks conveyed the atmosphere of "an Orwellian ritual-denunciation session."

In the twenty-five years since Jensen's paper, however, a shift in public sentiment brought about a more dispassionate hearing for *The Bell Curve*'s unpretty ideas than Jensen had received. In general, possible genetic explanations appeared far less preposterous. If the book's genetic explanations persuaded anyone, it would be sadly ironic timing in that geneticists were just beginning to discover mysterious ways in which repeated environmental stress can bring on permanent alterations in body chemistry, even changes in gene action. When decriers of the Murray-Herrnstein book spoke of the lingering effects of 250 years of slavery and oppression, it smacked of Lamarckism, but, as an explanation of the fifteen-point I.Q. gap, they may have been on firmer scientific ground—thanks to new research—than could have been believed a year earlier.

As with the Jensen paper, *The Bell Curve*'s critics, instead of focusing on the highly specific charges of unalterable I.Q. differences between the races, felt it necessary to wage war on the entire concept of genetic influence on behavior. They saw offshoots of behavioral genetics such as *The Bell Curve* as exposing the field's sinister political agenda. While some behavioral geneticists like Sandra Scarr said publicly that it was a good thing to have these sensitive matters aired, others, like me, deplored the book, knowing the critics would seize on this one racial application of gene-behavior theory as a club to beat the entire discipline back into shamed silence.

THE FINDINGS ALLUDED to above are bringing about a new understanding of the environment's subtle power to alter genetic expression and go well beyond such simple concepts as fertile or arid soil. In the functioning of an intricate, brainy organism like the human, a broad array of environmental elements are now seen to have important effects, sometimes lasting ones. A highly evocative example of the newly perceived genes-environment complexities involves I.Q., which most psychologists have finally come to agree has a high degree of genetic influence.

In Japan there is an ethnic group called the Burakumin that is shunned and disdained by other Japanese in much the same way as India's untouchables. Contributing to this opprobrium, the Burakumin consistently score lower on measures of intelligence and are widely considered a genetically inferior breed, although they look like other Japanese. This pat explanation turns out to be inaccurate. When members of the group move to California, where no one knows a Burakumin from an Osaka aristocrat, their children perform as well as other Japanese. This phenomenon carries the strong implication that assumptions of low I.Q. are, in some mysterious way, self-fulfilling appraisals, genetic makeup notwithstanding. Possible parallels with America's black population are unavoidable and may turn out to explain fully the consistently lower black I.Q. scores.

Research along these lines is currently being done by a prominent black psychologist, Claude Steele, who has found that black students do much better on tests if told the test's purpose is not to measure anything but to try out a new test format, or a new electronic pencil, or some other nonthreatening tactic. In subsequent runs, if the students were told that the test was to determine their intelligence, scores dropped dramatically. It appeared that the students' anxieties and mind-sets played a big part in their performance.

With the Burakumin, living as disdained "inferiors" placed them, in all likelihood, outside a normal range for proper function, perhaps for proper development. The black students' pretest briefings appeared to have the effect of either relaxing them so they could achieve maximum performance or crippling them with anxiety. While there is nothing particularly complex about these mechanisms, everyone knows that if you are nervous you generally perform less well. Until now, however, no one has thought of a large segment of the population being "nervous" for three or four generations. But that is what the Bukarumin evidence suggests.

In the two thousand years of swings between environmental and inheritance explanations of human behavior, we have arrived at a period when the genetic contributions are dominating public attention. At the same time, psychologists are discovering intricate, and hitherto unknown, ways in which the environment affects genetic expression. It would be an irony and a misfortune if, in our new enthusiasm for behavior's biological underpinnings, these new insights were overlooked, as the genetic contributions were for so many decades.

FIFTEEN

OH SO POLITICAL SCIENCE

WHEN THE ENVIRONMENTALISTS dominated American thought, their criticisms of the small band who clung to a biological perspective were based on scientific convictions. Although energized by their theory's compatibility with political progressivism, they were convinced that their researches proved them right. No one doubted that their science came first, the political ramifications second. The behaviorists felt they had discovered what made humans behave as they do; believers in instincts and genes were simply wrong.

As counterresearch began to erode the scientific underpinnings of its theories, cultural determinism was forced to yield more and more of its turf to sociobiology and behaviorism to behavioral genetics. With the two mainstays of environmentalism in retreat, a new breed of critics emerged to take up the fight against the biological perspective. These scientists wasted little energy defending behaviorism, which increasingly appeared a lost cause, and they offered no alternative theory of human nature, as had Watson and Skinner. Still, they were so alarmed by what they saw as the right-wing political consequences of the biological perspective that they dedicated themselves to attacking its every advance. The result has been an apples-and-oranges war of science versus ideology that smacks less of the nature-nurture debates of the 1920s and 1930s and more of the church-versus-science struggles of earlier centuries.

Being scientists themselves, these critics sought out potential flaws in their opponents' science—the assumptions, methodology, analyses, and so on—which is never difficult to do with any study of the complex human. In short order, however, they invariably moved to *ad hominem* attacks on what they were convinced were their opponents' hidden conservative agendas. For proof of their charge, they could not have asked for anything better than Jensen's paper on race differences. A few years later, Richard Herrnstein supplied more ammunition in an *Atlantic Monthly* article by escalating the bad genetic news from the blacks to the entire lower class (the poor more often than not have lower I.Q.s). The conservative swing of the Nixon administration, with Daniel Patrick Moynihan's prescription of "benign neglect" for the underclasses, gave the critics further reason to unmask the new genetic perspective as reflecting a conservative swing and nothing less than a return to the Social Darwinism of the last century. They saw a direct connection between the adoption and twin studies then in progress and the policy chambers of the later Nixon White House.

By the early 1980s, the furor provoked by Jensen had subsided and research results continued to mount in favor of a genes-behavior link. Word was emerging from Minneapolis of startling results from a separated-twins study, and the reports were not confined to obscure scientific journals but appeared in the *New York Times, Science,* and other influential publications. For the critics the time had clearly arrived for a forceful counterattack.

Three of the leading opponents of behavioral genetics collaborated on a book that set out to deconstruct the new science and reverse the biological tide. The book was *Not in Our Genes,* and the authors were three of the most vigilant critics of the genetic view: Richard Lewontin, a population geneticist at Harvard; the indefatigable Leon Kamin, who was then at Princeton's psychology department; and Steven Rose, a neurobiologist at England's Open University. Although the book had slight impact, it is worth examining as a compendium of the arguments and methods of the opponents of behavioral genetics, arguments that these critics, and their shrinking band of allies, continue to make despite repeated refutations.

Throughout the text the authors, with admirable candor, proclaim their Marxist perspective and their "commitment to . . . a more socially just—a socialist—society." Few pages go by without references to "dialectics," "bourgeois society," and "capitalist values." The authors' apparently feel their clean breast about their politics permitted wholesale assumptions about those of their opponents. We are leftists is their implicit claim; but you on the other side of the scientific fence are reactionaries. Liberals, they appeared to be saying, can have only one scientific view, theirs; any other must be right-wing and antiliberal.

"Biological determinist ideas," they say, "are part of the attempt to preserve the inequalities of our society and to shape human nature in its own image." It must surely have come as unpleasant news to Sandra Scarr, Jerome Kagan, and other liberal psychologists to learn that they were striving to preserve society's inequalities. In addition, the authors' nasty assumptions of their opponents' motives must have been an eye-opener to the hundreds of microbiologists, lab technicians, DNA scanners, rat-runners, statistical analysts, and all the others engaged in behavioral genetics research who learned from the book that they were going to work each day "to preserve the interests of the dominant class, gender, and race."

But the falsity of the authors' premise goes well beyond slandering a few individuals. Throughout the text, the writers deny the possibility that scientists could exist who place their curiosity about the world ahead of their political agendas. Lewontin, Kamin, and Rose deny as well the possibility of any man or woman, including themselves, separating science from politics. ("Science is not and cannot be above 'mere' politics.") They leave no room for the scientist who is so intrigued by new information, in this case gene-behavior discoveries, that he or she is oblivious to alleged political consequences. For the authors, all scientists who seek out biological influences on behavior, from Darwin to Robert Plomin, are willing servants of the status quo, if not promoters of a return to feudalism.

Fundamental to this accusation is their assumption that belief in a gene-behavior connection leads *unavoidably* to a position of political conservatism. To make this second *tour jeté* of logic, they prepare a soft landing by altering slightly the vocabulary of the behavioral geneticists, revising moderate expressions like "genetic influence" and "genetic component" into the far more alarming "genetic governance" and "biological determinism." By escalating the language in this way, they neatly push their opponents into indefensible positions that are held by no one. Not only does this altered language render the geneticists far easier to shoot down, but it also perpetuates the outdated fear that genes equal immutability. In fact, the straw man of immutability pops up on every page of *Not in Our Genes* and carries most of their argument.

For all the horror the authors have of genetic claims, they are not, they insist, of the school that believes in a zero biological component to behavior. In one of their most prolix sentences, they put forth what they *do* believe: "We must insist that a full understanding of the human condition demands an integration of the biological and the social in which neither is given primacy or ontological priority over the other but in which they are seen as being related in a dialectical manner, a manner that distinguishes epistemologically between levels of explanation relating to the individual and levels relating to the social without collapsing one onto the other or denying the existence of either."

Except for the "ontological priority" business, few behavioral geneticists would dispute the statement, assuming they went to the trouble to unravel it. The authors are claiming for themselves the same position of gene-environment interaction to which virtually all behavioral geneticists subscribe. In order to create a scientific difference to justify their combat, they saddle their opponents with an extreme belief in total genetic governance. The text soon makes clear, however, that the real difference is their belief that in our corrupt capitalist society, genetic differences between individuals are irrelevant when held up against the virulent environmental injustices.

Having aligned themselves scientifically with the new thinking of a gene-environment mix, they then devote considerable space to proving that the entire gene-environment interaction is too complex, too contingent, too fluctuating to ever nail down in any useful way. And they hammer away at research projects that have done precisely that, which is to say most of the sociobiological and behavioral genetics research of the past twenty years. They suggest, for instance, that kinship studies are meaningless, in that siblings and other relatives may resemble each other for environmental, as well as genetic, reasons. In this helpful way they deftly put their finger on the dynamic that kinship studies have, for twenty years, been successfully unraveling.

When they turn their sights on studies that compare identical with fraternal twins, both raised together, they quite remarkably fall back on the well-refuted argument that MZs are more alike than DZs because they are brought up to emphasize their similarities.* As one piece of evidence they cite a study that indicates MZs do homework together more frequently than DZs. While it would not be immediately apparent to most people that this collaboration would bring the twins' I.Q.s closer together, the most interesting revelation is that the authors have gone to the trouble of finding a study that suggests variation in MZ and DZ environments, while totally ignoring the numerous studies that prove that such variations, to the extent that they exist at all, have zero effect on twin development. The studies, which were reviewed at the end of chapter 9, were planned and executed well before publication of *Not in Our Genes*, primarily in response to these very criticisms— which had often been made, primarily by these very men—and were aimed specifically at the points they had repeatedly raised. The authors could have criticized the studies and picked holes in the design and methodology, but to pretend this counterresearch didn't exist was scientifically irresponsible and dishonest.

When the three authors arrive at the subject of separated twins, they omit any reference to the Minnesota Twin Study, which had been in progress for five years at the time of the publication of *Not in Our Genes* and which obviated most of the book's arguments. The authors chose to attack instead the earlier, less airtight reared-apart twin studies, targeting in particular the high concordance in I.Q. scores of separated MZs. Indignantly, they explain this nagging statistic as a result of recruitment bias, contacts between supposedly separated twins, and family connections between the allegedly diverse households.

Thomas Bouchard had been well aware of all of these criticisms of the earlier twin studies and designed his own study to meet them. Indeed, the elimination of such flaws had for him been a powerful motive in undertaking another separated-twin study. In the eyes of

*In addition to Kenneth Kendler's exhaustive refutation of this point summarized at the end of chapter 9, Robert Plomin cites five studies that show differences in treatment of MZ and DZs are slight and do not have *any* effect on personality and behavior. The five studies came out years before *Not in Our Genes*: P. T. Wilson, 1934; A. Lehtovaara, 1938; R. Zazzo, 1960; R. T. Smith, 1965; J. C. Loehlin and R. C. Nichols, 1976.

most observers, he was succeeding. He also seems to have been succeeding in the eyes of the *Not in Our Genes* authors because they
chose to ignore the largest and most extensive separated-twin study
that had ever been undertaken, while laboriously picking apart the
vulnerable previous studies.

One of the most telling examples of the way *Not in Our Genes*
ignores data concerns I.Q. findings. After attacking the earlier studies
on the usual (disproven) grounds, they make no mention of the Minnesota study, which avoided the very pitfalls the authors cite yet had
resoundingly replicated the earlier MZA studies' high I.Q. concordance figures.

When trying to fathom how eminent scientists could manipulate
their evidence in this unconscionable manner, it is perhaps useful to
remember that for some, political ideology takes strong precedence
over scientific facts. In a conversation I had with James Watson, the
codiscoverer of DNA's structure, he told me that physicist Salvadore
Luria had once said to him that "politics are more important than science." The casual statement of priorities suggests how in those of like
mind this might justify scientific skulduggery. To some, noble ends justify ignoble means.

While science is supposed to examine evidence and draw conclusions, the politically oriented start with their conclusions of how the
world should be, then bend or edit the evidence to fit that conclusion.
This was standard operating procedure in Russia under Stalin and in
Germany under Hitler, and it still thrives in American universities in
the guise of political correctness. In researching this book, I was quite
surprised to find the media that reports on scientific developments
seem unaware that this goes on. They are surely motivated by a desire
to be evenhanded, or perhaps they just want to stay out of the fray. For
whatever reason, they dutifully report the destructive obfuscations of
ideologues like Lewontin; but in so doing, they appear to me to be
allowing themselves to be used in ways that have nothing to do with
science. As a journalist, I feel that once a source has shown a willingness to lie for higher political goals, that source has disqualified him-
or herself from future input.

Oh So Political Science

FOR THE BEHAVIORAL GENETICISTS exposed to such politics-driven criticism, an added frustration is their opponents' failure to conduct their own research to substantiate their nihilistic arguments. Counter-research is, of course, the accepted method of resolving scientific disputes. If you feel a colleague's conclusions are wrong, you design and execute a study that will expose the error. The critics of behavioral genetics, on the other hand, confine themselves to the far less arduous course of attacking their opponents' research, finding weaknesses, possible flaws, and so on. As Arlen Price put it, "Kamin and the others sit back in their armchairs and take shots."

Except for Leon Kamin, most of the critics do not bother to inform themselves thoroughly about the research they are attacking. This sometimes results in their citing nonexistent flaws. A frequent criticism of separated-twin studies was that the twins had actually had contact during the so-called years of separation. While this was true, to a degree, with the earlier studies, it is definitely not true of the Minnesota study. Yet the critics continue to hint darkly of twin contact that undermined claims of "separated."

Jonathan Beckwith, a molecular biologist at Harvard and a political activist, is one of the more responsible and fair-minded of the behavioral genetics critics. Yet when I spoke with him, he mentioned this as a problem of the Minnesota study, citing a specific example, Jack and Oskar, the Hitler Youth and his kibbutzim brother. It was bad luck for Beckwith that he cited a pair of twins who were separated in the first six months of their life and did not see one another for twenty-one years, during which time they both lived in dramatically different environments.

"They were not really separated," Beckwith said authoritatively. "They had contact."

I pointed out that the first contact occurred when the twins were twenty-one, a fact ascertained by both twins, who had no reason to lie about it, and substantiated as well by Jack's ex-wife, hostile to her former husband, who was present at the reunion. Beckwith shifted his ground: "Well, they were both brought up in German households," he said. "Their upbringings were similar."

Aside from the odd supposition that two boys brought up in German households, *any* German households, would turn out with Jack

203

and Oskar's catalog of shared quirks and traits, the fact is Oskar's German household was his Christian grandmother's house in a village on the German-Czech border, while Jack's "German household" was a series of Jewish households—one in Trinidad, one in Venezuela, and two in Israel. If this would satisfy a scientific mind as explaining the Jack-Oskar similarities, it would suggest an even more staggering list of identical tics and idiosyncrasies that would inevitably be shared by any two boys growing up in different homes in, say, Berlin or Munich.

In a paper criticizing the Minnesota Twin Study, Beckwith and coauthors elaborate on the failings of Jack and Oskar as separated identical twins. With a smug tone of this-is-the-real-story, they point out that, after their meeting, *their wives corresponded.* When I interviewed Crystal Yufe, she was quite certain there had never been anything more than a few Christmas cards between households. Even if they had corresponded, it is remarkable that a scientist of Beckwith's standing would suggest letters between wives could result in a precise duplication of their husbands' oddities, personalities, and abilities. Beckwith goes on to suggest that the physical similarity in the two men could have resulted in similar treatment while growing up (in Germany and Trinidad). This ingenious, if slightly far-fetched, possibility has been previously tested empirically, as was mentioned at the end of chapter 9, and was found to have no effect.

In the paper, Beckwith and the others make the charge that separated-twin studies are based on the assumption that each twin's environment is markedly different. There is no such assumption. The studies assume that in measuring a group of reared-apart twins, their environments are on average *more* disparate than those of two twins growing up in the same house. The Beckwith paper also questions the assumption, which he falsely terms essential to twin studies, that twins growing up in the same household have the same environment (his doubt on this point puts him in the forefront of behavioral genetics thinking). On the preceding page, however, he made the opposite point by positing a reason that Jack and Oskar are so similar: They were both raised in "households with a traditional German character." Not the same household, but German, and this, he suggests, made the homes similar enough to bring about the twins' long list of parallels.

There is no particular reason why Beckwith, who is engaged in important molecular biological research unrelated to human behavior, should be well versed in the details of behavioral genetics research—unless, of course, he chooses to attack it. But attack it he and his co-authors do because of the dire consequences they see for liberalism in the implications of the Minnesota twin studies. While they may feel they help their cause with their half-baked attacks, tripping up by any means scientists with sinister motives, they are doing little for their own reputations for scientific thoroughness and accuracy.

Bouchard has frequently addressed the criticism that twin similarity results from placement in socioeconomically similar homes. Aside from the obvious rebuttal that fraternal twins raised in the same home are not as similar as reared-apart twins, Bouchard makes an additional point. "If similar home environments made people similar," he said, "close correlations would be found between twins and their adoptive parents (living in the same home). But a number of studies have shown that such correlations are not found."

Beckwith's errors about the twin studies are typical of the criticisms, often from men and women who are meticulous in their own science, and from opponents like Beckwith, who, in spite of their hostility to behavioral genetics, would not consciously fabricate discrediting information. When, however, they shoot from the hip this way, their improvised and inaccurate swipes are given credibility by scientific reputations derived from unrelated research. Their attacks are picked up by a press that struggles to tell both sides of controversies, and the cry of "fatal flaws" is heard throughout the land.

Bouchard is even more bothered that the critics segue quickly from pointing at weaknesses that *might* skew study results to claiming that these weaknesses *have* skewed results. And they do this, he points out, without offering any evidence that this is so. As mentioned above, they also do it in the face of abundant research data that have tested the premises and found them to have no effect. Yet the critics persist in the charge that these potential weaknesses undermine twin studies.

THE ACCUSATION OF FLAWS in research is a common weapon in scientific debate but has particularly plagued twin and adoption studies. The complexity of the study designs—their unwieldy numbers, varying subject ages, varying environments, varying lengths of separation—have made them relatively easy targets for critics bent on inflicting harm.

When, in spite of the lack of counterresearch, the critics refer darkly to "serious flaws," or the most lethal, "fatal flaws," these tidy phrases, so easy to put into the air, have a way of sticking, as does most destructive gossip. Sideline observers of the fray, even those with no stake in the matter, pick up the hurtful phrases and repeat them, unaware they are baseless. When in doubt about scientific matters, it is better to come down on the side of caution. As an added bonus, a pungent phrase like "fatal flaws" also lends a quick dose of self-promotion in that it places the speaker's standards above those being debunked. In this way, much injustice is perpetrated against meticulous and scrupulous researchers, and much significant data brushed aside.

Bouchard and other behavioral geneticists accept this as the inevitable price of working in a controversial field. They also see it as inevitable when studying a highly complex organism like the human, who does not fit into experiments as handily as mice and fruit flies. Some, like Minnesota's Matthew McGue, view this carping in a positive light. "Behavioral genetics gets people excited because it is important. Most things don't warrant any controversy because they just don't matter, but our field is considered significant enough to argue about."

Robert Plomin is another behavioral geneticist who accepts the criticism as coming with the territory but wishes it were not so. He wistfully cites an illustration of his notion of the proper way to wage scientific battle. The eminent British psychologist Sir Michael Rutter was convinced that autism was environmentally caused and was distressed by the growing tide of scientific opinion that held it to be genetic. To prove his hunch correct and halt the tide, Rutter set up his own experiment. The results, to his surprise, proved the opposite of his expectations: Autism showed a significant genetic basis. Quite publicly, Rutter made it clear he had changed his mind completely on the subject. Rutter's acquiescence to the evidence was, in Plomin's view, the only honorable response for a scientist, but one sadly lacking among the radical environmentalists.

Others in the behavioral genetics camp object to what they see as the unfairly high standards to which they are held as opposed to other branches of the social sciences. Sandra Scarr said, "Yes, we are complaining, not because of the final verdicts, but because the process is so manifestly unfair. Our work is subjected to the scrutiny of an electron microscope while the rest of psychology is examined through the wrong end of a telescope." Scarr continued: "Leon Kamin approves of the close scrutiny because he sees dire consequences in the possible errors of genetic claims. I am sympathetic with his politics in this case, but my civil-libertarian bias impels me to claim the right to make as many mistakes as anyone else in psychology."

Scarr is underestimating Kamin's opposition. He is not simply combating errors; he combats *any* evidence of a genetic role in human behavior. An examination of the indiscriminate ferocity of his attacks raises the question of whether he launches them because he believes there is no genetic role or because he feels such a belief is bad for business—that is, liberal politics. If pinned down, he acknowledges a genetic component to human behavior. Yes, he is saying, there may be a genetic component to behavior, but no, they haven't proved it yet. His rhetoric implies something quite different: that genes are in no way involved in behavior and all efforts to establish a connection are bogus.

Other beleaguered behavioral geneticists complain about the practice, when research is published, of publishing in the same issue vigorous rebuttals from one or more opponents. This too, they say, is a major deviation from normal scientific practice and appears an effort at preventing dangerous hereditarian positions to float unchallenged. When other claims of scientific advances are published, they are granted their moment in the sun—at least until the publication's next issue, when letter writers can do their discrediting worst.

The antigenes critics seek to deny behavioral geneticists even this degree of a hearing and do not always wait for papers to appear before launching their negative campaigns. Dean Hamer told me that when he was setting up his major study on genetic links to homosexuality, the gene police at Harvard got wind of it through a pamphlet Hamer issued to enlist volunteers. He received an immediate letter from biologist Ruth Hubbard, whose gene phobia was displayed in her 1970s efforts to organize Harvard office workers to demonstrate against recombinant DNA research. Hubbard told Hamer that his planned study was not the way to go about finding linkage (Hamer's scientific credentials are every bit as good as Hubbard's). A Harvard colleague, she told him, was in fact using the pamphlet in a course as an example of how *not* to conduct linkage studies.

Hamer next got a letter from the colleague, Evan Balaban, asking

Hamer for methodological details about his study. With less candor than Hubbard, Balaban said that his aim was to survey "conceptual advances" in behavioral genetics research. Still smarting from being "blind-sided by some Harvard professor I never met," Hamer received yet another letter from Richard Lewontin, who was co-teaching the stamp-out-genes course with Balaban and was, he said, writing at the suggestion of Ruth Hubbard.

Like Hubbard, Lewontin proclaimed his opposition to the methodology of Hamer's study, stating that *behavior* was "very, very far from genes." He cited language as an example of a behavior that is *not* influenced by genes, pointing to the world's diversity of languages, which "we know for sure are not influenced by genes." (This would surely come as jarring news to Noam Chomsky, Stephen Pinker, Myrna Gopnick, and other experts on language theory.) Lewontin also cited the failed effort to locate specific genes for various behaviors (alcoholism, for one) as reason to abandon this line of investigation.

Hamer was stunned that prominent scientists were ganging up on a study that had not yet begun. More than attacking it, they were trying to stop it. When I met with Hamer in Washington, he expressed amusement at the disingenuousness of Balaban's feigned interest in the gay-gene quest,* but was less amused by Lewontin's condescending lecture on the distance between genes and behavior. In responding to Lewontin's point about diverse languages, Hamer said, "That's like saying that the urge to eat is not genetic because some people like hot dogs and other people prefer tacos."

When he realized that the Cambridge group's intentions were less friendly than Balaban had indicated, Hamer volunteered to give a guest lecture on his DNA study to the Harvard class, hoping to call their bluff about interest in his methods. To his surprise, they accepted

*Ironically, this same Evan Balaban won headlines in 1997 with an experiment that demonstrated, perhaps more vividly than had Hamer, the direct link between genes and behavior. Working at the San Diego Neurosciences Institute, Balaban planted brain cells of quail into the brains of embryo chickens. The result was chicks that did not sound off like chickens but crowed and bobbed like quail. Remaining true to the faith of his Harvard fathers, Balaban dismissed the implications for humans of this landmark behavioral transplant, announcing that "most human behavior is learned" *(Time,* March 17, 1997). Alfred Wallace and Bishop Wilberforce must have smiled down at this reaffirmation of human uniqueness.

his offer. At the lecture, Hamer outlined in careful detail what he saw as the criteria for a sound behavioral genetics study, stressing the statistical validity of teasing out genetic influence on complex behaviors by comparing the DNA of siblings who share the behavior. In his fascinating 1994 book on the entire gene-hunt adventure, *The Science of Desire*, Hamer describes the scene at his lecture's end:

> When I finished, I asked for questions, bracing myself for what I thought would be a barrage of hostile criticism, perhaps a few hoots of derision, and a generally unpleasant thirty minutes. Instead, Lewontin stood up and graciously announced to the class that I had addressed essentially all the concerns that he and Balaban had about human behavioral genetics studies. As far as he was concerned, Lewontin said, our study was scientifically sound. I had been forgiven. I wasn't sure if I should kiss Lewontin's ring, but he ducked out of the lecture hall without speaking directly to me.

No doubt Lewontin was hurrying off to find other behavioral genetics studies to attack, ones more vulnerable than Hamer's. When he got to his Boston hotel room that evening, Hamer learned that his Maryland computers had indeed found what they had been looking for: the linkage of homosexuality to a downstream region of Xq27 with a 3.0 LOD score (a logarithim of the odds). This meant there was only one chance in a thousand the result was not statistically significant.

WITH OTHER EPISODES LIKE Hamer's, many behavioral geneticists were concluding that the critics were not interested in serious scientific debate. Thomas Bouchard tries to see a more positive aspect to the relentless criticism than mere validation of their work's importance. He sees it as guaranteeing that he and other behavioral geneticists will strive for the highest, most painstaking, standards in their research. When setting up studies, they are all too aware of the denigrators who lie in wait to pounce on the most trivial potential flaw. The critics, of course, claim that keeping the science pure is their only motive in carping, but even a casual examination of their complaints shows they are intent on discrediting this science in any way they can.

Most behavioral geneticists know that even their most airtight data will be attacked. If their data withstands the first onslaught, they brace for the character assassinations that usually come next. The geneticists proceed in the hope that more reasonable, less politically motivated scientists will see through the critical smoke screen to the research beyond—as indeed happened when influential scientists from other fields, such as *Science* magazine's former editor, Daniel Koshland, examined the studies and found them valid. They are also encouraged by the honesty of fellow psychologists, such as Jerome Kagan and Sir Michael Rutter, who were forced to accept the unwanted implications of their own research.

It has always been a fact of scientific life that the ease of acceptance of new information depends in large measure on its compatibility to the day's political mood. Little hard scrutiny was given the theory-serving experiments of Watson and Skinner; when their claims were eventually subjected to empirical tests, they were found wanting. It is also illuminating to contrast the nitpicking examination of behavioral genetics studies to the free ride given Sigmund Freud. Although his extravagant theories were never tested, they still revolutionized psychiatry and overhauled the way humans viewed themselves. Freud achieved all this with vague claims of "clinical observation," details of which he refused to divulge, even to his staunchest supporters in the early years. Considering the very different treatment given behavioral genetics by the rest of science, Sandra Scarr, in a philosophical mood, said, with admirable understatement, "Standards of evidence are raised when one is defying the zeitgeist."

The attacks on behavioral genetics research, rash and uninformed as they often are, flew completely out of control in one instance and provided a chilling glimpse of the ferocity of the antigenes forces. In Jensen's incendiary paper on I.Q. differences, he had relied to a degree on the separated-twin research of eminent British psychologist Sir Cyril Burt, who died in 1971 at the age of eighty-nine. Considered by many to be the father of modern British psychology, Burt had reached a level of influence and power rarely achieved by a scientist. Among other posts, he sat on the London County Council and held sway over England's educational policies in the turbulent postwar years. As might be expected, his research on the heritability of I.Q. informed the policies he advocated.

It is ironic that Arthur Jensen, who agreed with Burt about I.Q.'s heritability, was one of two who first cast doubt on Burt's research. The other whistle-blower, more predictably, was Leon Kamin. One of the main complaints of Jensen and Kamin was that Burt's concordance figure for I.Q. in separated twins remained a constant .77 through a number of different studies involving numbers of twin sets. The chances of the number repeatedly coming out precisely the same were nil. Both men were also bothered by Burt's having apparently *estimated* the I.Q.s of some twin parents, yet treating these estimates as hard scientific data. These concerns about Burt's science led to a bizarre drama that was perhaps the blood-lust apex of the nature-nurture wars.

The allegations against Burt's work were picked up and augmented by two English psychologists, Ann and Alan Clarke of Hull University, who were former students of Burt's. The Clarkes raised the possibility of fraud. Others joined in and eventually the rumbles of a major scientific scandal were overheard by an investigative reporter at London's *Sunday Times*. The newsman examined Burt's research and found other curiosities, such as a dearth of data about Burt's twins and his claim that World War II bombings had destroyed many of his records. When the reporter was unable to find evidence of the existence of two research assistants Burt claimed to have had, he decided to go ahead with a major exposé. It appeared in the *Sunday Times* in October 1975. England's most distinguished psychologist was denounced as a mountebank who had lied about his work and falsified data.

It would be hard to overestimate the furor that swept the scientific community. Most found it inconceivable that such an eminent scientist, a man upon whom every honor had been thrust, was in reality "a confidence trickster," as the Clarkes termed him. The many distinguished scholars who came forward to defend Burt could offer little more than character references; none could provide satisfying answers to the many questions about Burt's decades-old research, and, even more suspect, none could verify the existence of the two phantom assistants. The attackers finally prevailed, and the defenders backed away. The chorus of denigration climaxed with Burt's biographer,

Professor Leslie Hearnshaw of the University of Liverpool. In his 1979 book *Cyril Burt, Psychologist*, he sadly agreed that Burt had fudged his data. This was the clincher that led the British Psychology Society formally to accept the guilty verdict about one of its most illustrious members.

For the next twenty years Cyril Burt served as the poster boy for scientific fraud; any references to his work were contemptuous dismissals designed to prevent gullible young psychologists from taking his research seriously. But hints began to emerge that perhaps Burt had been misjudged. Finally, two books appeared, one in 1989 (*The Burt Affair* by Robert B. Joynson) and the other in 1991 (*Science, Ideology and the Media* by Ronald Fletcher), that set out to show that the charges against Burt were unfounded or exaggerated, that at worst he was guilty of careless methods, or more precisely, methods considered careless by the standards of the later period in which he was judged. Neither book pretended to answer all questions about Burt's research, but both were convincing that the evidence was insufficient for the knockout charge of willful fraud.

Both authors have little trouble disposing of one of the most incriminating pieces of evidence against Burt, the identical correlation numbers for I.Q. in different studies involving different numbers of twin sets. Investigation indicated that Burt had not retested in the particular category and had simply used the figure from his earlier tests. This is dubious practice and should have been announced by Burt; but without further evidence, it cannot be termed fraudulent. Both writers, with sound logic, point out that anyone deliberately inventing data would take the obvious precaution of altering the numbers slightly to throw off suspicion. Since this ruse is known to anyone who has ever estimated tax deductions or filled holes in an expense account, it would surely have occurred to one of England's finest minds, bent on a higher order of chicanery.

The books offer no help in unraveling the mystery of the perhaps nonexistent research assistants, which lent an Agatha Christie air to the scandal. The best each author could do was track down and interview associates of Burt who had vague recollections of the women but could not swear to their names. In the two books, this was insufficient to lift the charge that he invented the assistants. (Since publication of the exonerating books, one of the women was found, a Miss Conroy, who had emigrated to Australia.)

Both books succeed in their main purpose of establishing that Burt's reputation was destroyed on too little evidence, and both make fascinating reading as scientific detective stories. Their chronological presentation of the steps leading to Burt's disgrace offer a terrifying view of a burgeoning lynch-party mentality. Numbers of journalists and distinguished psychologists were like schoolchildren circling a victim, each pushing their taunts a bit further than the one before, all waiting to see who would muster the courage to strike the first real blow. London's *Sunday Times* proved the bravest.

Although the actual charges of fraud against Burt came mostly from the English scientists and journalists, it was Kamin who had submitted Burt's research to a thorough review and declared it worthless. While this is a devastating judgment on a colleague's life work, scientists, especially Kamin, do this to each other with nerve-wracking frequency. When such a charge was leveled at the august Burt, however, it opened the door to the crescendo of escalating accusations that took place in England. Although Kamin had only raised the *possibility* of fraud, one can only imagine the satisfaction with which he watched from across the Atlantic the pillorying he had set into motion.

One of the authors defending Burt, Ronald Fletcher, subjects Kamin's critique of Burt to analysis and finds many inaccuracies and omissions, all on the side of Burt's purported fraudulence. Fletcher documents this skewed critiquing so thoroughly, in fact, that it raises the suspicion that Kamin felt confident no one would trouble to check him out as he was checking out Burt. However deliberate the distortions in Kamin's analysis may have been, it was widely accepted and carried considerable weight. It provided what many wanted to hear: I.Q. still had not been proven to be heritable. It also provided the tasty subtext that those who said it was were scoundrels and con men (which is the central theme of Stephen Jay Gould's book on I.Q. testing, *The Mismeasure of Man*).

Burt based his claim of I.Q. heritability on a reared-apart twin correlation of .77. When the attacks on Burt began, his correlation figure had already been corroborated by three earlier studies and has been replicated by two subsequent ones—Bouchard's in Minnesota and the

Swedish study of identical twins. Although Burt's figures clearly contained irregularities and puzzles, his bottom-line conclusion has been confirmed, not only by the two Burt-scandal authors, but also by a wealth of unrelated research. This wholesale substantiation only adds to the irony of so many people expending so much energy in a wild-dog effort to destroy Burt. It is even more ironic that for twenty years they were successful.

GIVEN BURT'S GRISLY FATE, it is surprising that anyone, Swedish or American, would venture into the shark-infested waters of I.Q. heritability. In fact, when Bouchard set up his twin study, the mangled reputation of Cyril Burt hung heavily in his mind, with so much of the doubt based on the *absence* of documentation, not its falsification. As a result, Bouchard took stringent measures to ensure that every aspect of the Minnesota study was extremely well documented—recruitment of twins, correspondence, interview tapes, medical test results—all the accumulated data of each twin's involvement with the study was meticulously recorded and stored. Perhaps the most poignant example of Bouchard's obsessive documentation was his videotaping of his interviews with each twin set.

"I had no specific scientific purpose in mind," he later said, "but I didn't want them coming along years later saying I never had any twins, that I made the whole thing up." As "they" had with Cyril Burt, he might have added. Since Bouchard is not a man of exaggerated fears or paranoia, his belief in the need for hard evidence that the Minnesota Twin Study had actually taken place provides a chilling picture of the ruthless opposition he felt himself to be facing.

Some two dozen scientists had been involved in Bouchard's study at varying times, as well as hundreds of twins. The project was well known at the University of Minnesota and around the country, his laboratory had been visited by many journalists and scientists, among them Leon Kamin himself, and his twin study had been written up in numerous publications from Sunday supplements to the nation's most prestigious scientific journals. Still, Bouchard felt that without actual videotape of each twin set, he was vulnerable to charges of fraud. Anyone familiar with the Cyril Burt case could not scoff at Bouchard's fears.

In fact, a complaint already made against the Minnesota study had implications of fraudulence. The critics spoke often about the unavailability of specific data about Bouchard's twins, although knowing that releasing such data is forbidden by federal law. On agreeing to take part in the study, the twins signed an informed-consent agreement in which the Minnesota scientists promised not to divulge information that could be traced to specific twin volunteers. Bouchard points out that under these rules, Juels-Neilsen and other twin researchers could not have published as they did; too much of their information could have been traced to particular twins. Even though the critics have been repeatedly informed of this restriction, they complained noisily about Bouchard's refusal to divulge specific twin data. This evokes the specter that Bouchard and his sixteen coconspirators were falsifying data.

Such smear innuendoes are a tradition among the antigenes group. In spite of Konrad Lorenz's Nobel Prize for his animal studies, the opposition constantly alleges that he cooperated with the Nazis when a young scientist working under their regime. In the same McCarthyite spirit, the critics point to Minnesota's acceptance of funds from the politically suspect Pioneer Foundation as evidence of their evil intentions. Bouchard's arrest for student radicalism, on the other hand, is never mentioned. The critics are as selective in their use of *ad hominem* data as they are with scientific data.

When Ruth Hubbard accused Dean Hamer, before he had begun his study, of not only bad science but bad politics, she came closer than most of the environmentalists to admitting that their arguments rested on the belief that certain knowledge is "bad" and should be avoided or ignored. Most scientists take the position that knowledge is neutral, value free; the use to which it is put might be good or bad, beneficial or hurtful to society in general. First learn as much as we can, then let society decide how new information will be used. The opponents of behavioral genetics have consistently feared such a climate of unfettered inquiry.

SIXTEEN

SCIENTISTS IN DENIAL

NOT ALL OF THE CRITICS of behavioral genetics resort to attacks on the integrity of the scientists involved. Some restrict themselves to scientific argument and, because of this, attract less attention than the character assassins. Prominent among this more temperate group is Jerry Hirsch, a professor of psychology at the University of Illinois. Unlike many psychologists, Hirsch is well grounded in gene theory and is considered one of the founders of behavioral genetics.

When the furor over Arthur Jensen's paper erupted, Hirsch was as appalled as the environmentalists and turned his knowledge of the new science against Jensen and his methods. Disagreement between members of the same discipline happens constantly, but Hirsch's outrage at what he saw as Jensen's malevolent misapplication of behavioral genetics sent him into a frenzy of contradiction, from which he never fully recovered. For a good part of the years that followed, Hirsch devoted considerable energy to castigating what he saw as the excesses and oversimplifications of behavioral genetics.

His fundamental argument is that, with humans, there is too much variation in both environment and genetic makeup for any coherent study of gene-environment effects. In the epilogue to a book he edited, *Behavior Genetic Analysis*, Hirsch writes, "Since genotypic diversity and genotype-environmental influence are ubiquitous, attempts to study the laws of environmental influence have been grasping at shadows." Having worked extensively with fruit flies, Hirsch developed a two-tiered view of the problem facing behavioral genetics: the ease of controlling fly environments as compared to human, and the enormous potential for genetic variation and anomalies in any organism. Looking at the messy, unmanageable human from this vantage point, Hirsch threw up his hands at ever sorting out the elements with any scientific accuracy. He also saw little use to heritability figures.

A student of Hirsch, Tim Tully, who now researches learning and memory in fruit flies at Cold Spring Harbor Laboratories, is a forceful exponent of Hirsch's views. Tully dismisses twin and adoption studies for two seemingly contradictory reasons. Like Hirsch, he feels human environments are too variable and their effects too difficult to measure. Environmental elements like diet, climate, and so on cannot be controlled, as they can be with fruit flies. On the other hand, Tully feels the human samples are not broad enough; there is not *enough* variation in the environments of the people studied, who are mostly white, middle-class Americans—no African bushmen or Peruvian Indians, and, most politically relevant, no black Americans from the inner cities.

To demonstrate the environment's power over genes, Tully cites a 1907 experiment by Stockard in which magnesium salts were added to sea water in increasing amounts until two-eyed minnows gave birth to one-eyed offspring. "The idea of one nose, two eyes is as genetically determined as anything," Tully said to me when I visited his Cold Spring office. "What could be more ubiquitous than two-eyedness? But Stockard showed that if you raise the magnesium level in sea water, the fish embryo would develop one cyclopean eye . . . It didn't survive. But there you go, the genetic assumption of two-eyedness disappeared when you change the environment."

While Tully was clearly citing the ninety-year-old experiment to demonstrate how susceptible to environmental influence the most rigid genetic instructions are, I later thought it could be seen as proving the opposite. When you consider how many two-eyed creatures there are in the world and, within a species such as *Homo sapiens*, with populations thriving in environments that range from Amazonian jungles to Siberian ice fields, and when you consider how many of them emerge with two eyes—rarely one eye and never three—it speaks more to the determination of genes to express themselves in spite of environmental fluctuations.

The world abounds with examples of the awesome power of genes to do their thing in spite of impossible conditions. Outside my office in Key West, where I am writing this chapter, is a brick barbecue pit installed by the previous owners. In the chimney, some five feet above the earth, an oyster plant has sprouted, finding enough nutrients in the cracked grouting to send forth its full complement of well-formed purple and green leaves. It sits there happily, a living chimney corsage, testimony to the oyster plant genome's determination to express itself, making do with masonry when no soil is available.

Of course this tropical plant would not thrive at the Arctic Circle or on the roof of a subway car. Any number of environments would defeat its urge to become. But to hypothesize such far-out conditions to prove gene malleability is self-defeating. That Tully and the other environment boosters must resort to extremes like magnesium-laced water to demonstrate the alterability of gene expression does no damage at all to the "normal range of environment" concept. It is a bit like saying that if you squash a bug under your heel, you are affecting the bug's genetic expression and proving that genes are readily altered.

(I found it interesting that even Tully, who confined his attack to the science involved, could not resist a swipe at their characters. After listing what he considered behavioral genetics fallacy after fallacy, he said, "You get to the point where you must ask, how many *ad rems* equal an *ad hominem?*")

The evidence is strong that whether it is magnesium-loaded fish or thalidomide human babies, it takes a lot to throw the gene locomotive off its track. Even the experiments with rhesus monkeys mentioned in chapter 9 in which repeated stressful experiences appeared to permanently alter the monkeys' neurochemistry, many would argue that this experiment places those monkeys outside of the normal range of environments.

With a firm grasp of the normal-range concept and an appreciation of the partial nature of genetic influence on behavior, the conclusion seems to be that, yes, genes are powerful in expressing themselves in developed organisms; no, they do not necessarily have great power over the organism's behavior. These are two distinct levels of potency. One is the level of power to express whatever the genome dictates concerning a behavior, which might be a weak or strong influence. The other is the level of power of those genetic expressions to affect that behavior. The two influences are quite different, yet, sadly, confusingly for the overall argument, they are often taken as the same.

An analogy frequently used by the environmentalists concerns corn seed raised in rich, fertile soil as opposed to the same seeds grown in dry, nutrient-poor soil. The first will grow straight and full, while the latter will be spindly and twisted. I could not see that such experiments, whether corn or minnows, did more than define the limits of that species' normal environmental range. No one argues that the environment cannot affect genetic expression. The question is, at what point?

Sandra Scarr and Thomas Bouchard are among growing numbers of psychologists who believe that, with humans, significant environmental alteration of genetic expression is relatively rare. Hirsch, Tully, Richard Lewontin, and Ruth Hubbard feel it goes on constantly. "We are not just read-outs of our genetic instructions," Hubbard wrote with exasperation in a letter to *The New Yorker*. Behavioral geneticists would answer that to a larger degree than ever before suspected, we are. They would also add that the environment—which accounts for roughly 50 percent of behavioral variation—is not just such well-known factors as rearing and education, but less malleable ones such as womb conditions and hormonal surges. To the degree genes do mold behavior and personality traits, their expression still allows for a wide range of performance and behavioral variation. Best of all for the genephobes, this variation would include a capability of totally overruling the ongoing genetic pulls and nudges.

In response to the critics' lament that the intricacies of gene-environment interaction invalidate efforts to sort them out, Minnesota's Matthew McGue responds, "That's tantamount to saying 'Let's not study human behavior. It's too complex.'" Jerry Hirsch, in effect, *does* say that. In a bow to this position, Bouchard admits that his studies have not challenged the breadth of possible interactions, in that his twins' environmental range is relatively narrow. "Our results are contingent on a reasonable range of environments and opportunities," he said. "It doesn't mean environment is irrelevant." But on other occasions he echoes McGue's exasperation with the Tully-Hirsch argument that human environments cannot be controlled, as can those of laboratory animals. "We can't control the weather or the stars," he said, "but people study them."

Anthropologist Robin Fox, who is now a professor of social theory at Rutgers, in his 1994 book *The Challenge of Anthropology* shows none of Bouchard's restraint in responding to the perpetual critical heckling. He dubs the opponents of the biological perspective "left-over, anti-system, left-liberal, chic-radical campus rebels and lumpen Marxists of the 1960s and 1970s . . . who have lazy minds." The final zinger refers to Fox's accusation that the critics have given up on the arduous task of understanding human nature and are content to mouth formulaic opinions instead.

The codiscoverer of the double helix, James Watson, has a more succinct summation of the Cambridge radical environmentalists. After reading a *New York Times* op-ed piece by one of them warning against drawing any conclusions from a major study that showed a genetic link with a common behavioral form, a piece Watson found distorted to the point of dishonesty, he snapped, "They're crooks!"

THE AUTHORS OF *Not in Your Genes* are not the only ones to saddle behavioral geneticists with claims they do not make or to use language that subtly misrepresents their work. I was to discover that it goes on constantly. In *Myths of Gender* by prominent feminist Ann Fausto-Sterling, she speaks scornfully about the "enormous appeal" of the view that "genes dictate our behavior." Aside from the highly arguable assertion of the appeal of a genetic view with all its frightening intimations of immutability, virtually no behavioral geneticist believes that genes "dictate" behavior. With a similar raising of the other player's ante by altering the words used, Ruth Hubbard, in her 1993 book *Exploding the Gene Myth* (written in collaboration with her son, Elijah Wald) complains, "The myth of the all-powerful gene is based on flawed science that discounts the environmental context in which we and our genes exist."

Anyone wishing to join Hubbard in the flaw-hurling sport could point to two flaws in Hubbard's sentence. "The myth of the all-powerful gene" is in itself a myth and can be found nowhere in the behavioral genetics literature. Neither would Hubbard find it easy to locate a behavioral geneticist who "discounts" the environment. Perhaps her misstatement of her opponents' position stems from the decades when Hubbard's team were on top and they routinely "discounted" genes and insisted on an "all-powerful environment." Now that the other side is in the ascendancy, Hubbard wrongfully attributes to them a similar extremism.

Word games abound among the critics. In 1976, when the I.Q. debate triggered by Jensen was still raging, Leon Kamin wrote: "The adopted child studies, like the separated-twin studies, seem to me to offer no evidence sufficient to reject the hypothesis of zero heritability of I.Q. scores." A number of years later I was interviewing him in his office at Northeastern University, where he had moved from Princeton. He is a slight man, bankerish, with an amiability and gentleness that belie the ferocity of his combat against the biological perspective. It was hard to imagine him as the man who had won the shoot-out at the O.K. Corral with Sir Cyril Burt. Finding him open and chatty, I asked if he really believed there was no inherited component to I.Q. Kamin smiled mischievously. "Of course there is a genetic component," he said, "but our techniques of measuring it are too flawed."

Few people aware of the volumes of evidence of I.Q. heritability could read his disdainful, one-sentence dismissal of it all without getting the impression that Kamin's position was: I.Q. is in no way inherited. Not crooked exactly, but not on the level either.

Kamin styles himself as the watchdog of behavioral genetics, the moral guardian of scientific probity. But this sort of verbal legerdemain, his omission of the Minnesota Twin Study when he evaluated all twin studies, his repeated use of arguments he knows have been disproven, and his highly slanted analysis of Burt's research all suggest he is playing a far more active role than merely keeping everyone else honest. While he is indefatigable in combing the behavioral genetics data for flaws, more thorough certainly than his fellow critics, he still has yet to do counterresearch, to set up studies that would disprove the work of Bouchard and others. For most scientists, this is the accepted way to refute and, in the end, the only one that means anything.

IN THE USE OF LANGUAGE, the radical environmentalists have even more invidious ways to misrepresent and trivialize the achievements of behavioral genetics. The champion in this sport is Richard Lewontin, one of the *Not in Your Genes* authors, who has repeatedly proved he needs no collaborators in his campaign of distortion. A stand-out example was in a review he wrote of a number of books on female biology for the *New York Review of Books*. After sneeringly discussing the advances in genetic therapies for various major ailments, he then turns to behavior. "The genetic model of disease leads ineluctably to the disease model of all ills and social deviance. So genetic defects are claimed to lie at the basis of heart disease, schizophrenia, alcoholism, drug dependence, violent behavior, unconventional sex, and shoplifting."

Behind the intellectually reassuring word *ineluctably* Lewontin hides twenty years of behavioral genetics research, of which he is well aware but which he chooses to keep from his readers. Instead we are left to suppose that anyone believing in a genetic component to alcoholism or homosexuality does so having leapt with wild illogicality from a belief that cancer and heart disease are genetic as well. And even these medical conditions are, in Lewontin's presentation, merely "claimed" to have genetic origins. No mention of studies, no pedigrees, no research, no reams of evidence—merely some overly excited geneticists have *claimed* these illnesses for their domain. Another relevant fact Lewontin fails to include in this article: *He* has a better cure for all civilization's ills and social deviance, a political one. Lewontin's ongoing performance as a leader of the antigenes crowd comes much closer than any to Watson's blunt characterization.

If the critics of behavioral genetics would limit their charges to the science rather than the personalities, and, within the science, if they would argue against what has actually been postulated instead of what they fear lies behind the postulations, or what they know will alarm neutral bystanders, they might have plausible arguments, but by ascribing to their opponents claims no one has made, they are tilting at windmills. In other cases, they cite the opposition's conclusions without giving the evidence on which those conclusions are based. They seem to believe that by leaving out any basis for such conclusions, others will see the achievements of behavioral genetics as windmills as well.

Because of this sort of twisting of the opposition's language, the entire debate is thrown into a kind of Looney Tunes melee when all the behavioral geneticists can do at the accusations is sputter, "Who?

Me?" In spite of repeated disavowals and efforts to set the record straight, the distortions and misrepresentations continue unabated and render understandable a cri de coeur Sandra Scarr made to me when I told her about a typically distorting letter to the *New York Times* by a Harvard genetics critic. "I wish," said Scarr, "they would shut up and listen to what we are saying."

IN SPITE OF THE VIGOROUS, sometimes pernicious efforts of the environmental holdouts, there has been a growing acceptance, both in the academic communities and with the public, of the concept of a genetic influence on behavior. The 1992 centennial conference of the American Psychological Association chose two themes that best represented the past, the present, and the future of psychology. Behavioral genetics was one of the themes chosen and a ten-hour series of symposia was organized to examine genetic research in psychology. With such impressive validation, the mainstream media, understandably shy on the subject that had been so politicized, began running interested, even enthusiastic, reviews of behavioral genetic discoveries. A culmination of sorts was reached in the summer of 1994, when a *Time* magazine cover said in bold letters (no picture): INFIDELITY: IT MAY BE IN OUR GENES. Six months earlier, Minnesota's David Lykken remarked to me, "The genetic component to behavior has now been well established. It's time to move on to other things."

Lykken's statement was premature. The June 1993 issue of *Scientific American* contained an article that purported to be an overview of behavioral genetics but was in actuality an inflammatory and deliberate misrepresentation of the field, with the clear intention of a wholesale debunking. The article's title and its cover line set the bias level. It was called "Trends in Behavioral Genetics: Eugenics Revisited." The use of the term *eugenics*, the most dreaded application of genetic theory, introduced the calumnious charge that behavioral genetics, the effort to find the genetic component in behavior, leads inevitably to social policies aimed at controlling reproductive rights. It was like saying chemists are ipso facto advocates of chemical warfare. The subtitle on the cover was: *The Dubious Link Between Genes and Behavior.* Had the editors specified *human* behavior, the magazine might have retained some degree of intellectual respectability (slight, since that would imply humans are not animals), but as it stood, the cover line proudly proclaims the editors' ignorance of two thousand years of breeding behavioral traits into animals.

In the text itself, written by a nonscientist named John Horgan, things get worse. In dealing with the Minnesota Twin Study, Horgan talks about the "weird" and "bewitching" coincidences and makes no mentions of the forty-odd research papers the study had produced over twelve years, only two of which allude to the "coincidences." While the scientific impact of the Minnesota Twin Study is based entirely on these papers, not on the coincidences, readers of Horgan's article are left with the opposite impression, that Bouchard and the others are basing their claims of genetic behavioral influences on a few odd twin similarities.

Readers can infer that there is more to it than Horgan admits when he resurrects all of Leon Kamin's long-since discredited criticisms of twin studies. Although Horgan was provided with the studies that refute the points (as Bouchard stated in a letter to the editors), he totally ignored them. All of the questions Kamin raises—does contact between twins make them more similar, do twins become more similar if, during rearing, their similarities have been stressed, does recruitment through the media attract twins who exaggerate their similarities?—all these points are matters that can be tested empirically. They have been tested and found to have no effect. But Horgan simply repeats the criticisms and leaves them hanging, as if no one had ever before addressed them.

Horgan states that behavioral genetics findings have not been replicated, even though Bouchard had explicitly told him of replications, most significantly the replication of his I.Q. results by the Swedish twin study. As it happened, the Swedish study used totally different methods to recruit their twins than had Minnesota, thus strengthening the replication of heritability figures and demolishing the Horgan/Kamin chestnut about media recruitment. In spite of having been informed of all this, Horgan writes that twin studies are unreplicated.

When Horgan turns to the DNA studies, he cites the misfires such as the Amish and Israeli studies on depression and the Blum study on alcoholism but makes no mention of the ten years of extensive observational studies that have established beyond question a genetic basis for some forms of depression and alcoholism. Here again the reader is left with a totally false impression, that the molecular biologists had gone off on a fool's quest for a left-handed depression gene based on nothing more than a hunch and had failed to find it.

With Horgan's presentation, the question of genetic links is still open, but the strong implication is that because of the failures to locate the specific genes, they aren't there. Horgan manipulated his data to suggest that molecular biologists were chasing phantoms, that the misfires indicated that there are no behavioral genes. Three misfires and you're out, he seems to be saying. Give up the quest. In reality, all that the early failures proved was that finding causative neucleotide bases among the three billion base pairs is difficult.

I went to see Horgan and confronted him about all these omissions and distortions. A handsome, dark-bearded man in his thirties, Horgan replied, "What's wrong with writing a polemical article?"

Understandably, the piece caused a furor and prompted more mail than any article in *Scientific American*'s history. According to Horgan, the letters were split roughly in half, pro and con. While I was in his office, he allowed me to look at the fat letters file. My casual perusal gave an impression that those applauding the article were of the thank-God-somebody's-halting-all-this-talk-about-genes variety. Many of the letters protesting the article were from scientists and academics, and their tone was one of outrage.

One of the few letters published was signed by Bouchard and fifteen other scientists from eight universities. They cited a number of Horgan's errors and distortions and protested stunningly false generalizations such as "practically every claim of a genetic basis can be explained as an environmental effect." The writers of the group letter, after charging that the "small coterie of critics" have never executed any research that would support their environmental positions, say, "For twenty years, [the critics'] main scientific strategy has been one of idle yearning, yearning for the body of findings to be exposed as specious or fraudulent or, increasingly, yearning for the facts to go away." When addressing Horgan's impugning of their motives as politically suspect and his allusions to Hitler, they write that such slurs "are merely an attempt to win with scare tactics that which has not been won in the laboratory."

Many unpublished letters came from scientists who had little connection with behavioral genetic research. Lee Silver, a professor of genetics at Princeton, started his letter with "I am shocked and dismayed that a journal as reputable as *Scientific American* could stoop to the depths of a mass-market magazine and allow the stand-alone publication of the extremely one-sided and highly biased article by John Horgan. . . ." Michael Levin, a professor of philosophy at the City College of New York, wrote, "Horgan repeatedly invokes eugenics and, to his eternal shame, the Nazis. Surely it is clear that what Hitler did has no bearing on the causes of observable human differences (nor, for that matter, on the propriety of studying the question)." David Rowe, a psychologist at the University of Arizona, protested many of the article's points, then concluded, "I hope that *Scientific American*'s next article on this topic will explore it with greater depth, thoughtfulness and veracity." In a conversation, Robert Plomin came out and said what many were thinking: "Bouchard told Horgan about scientific papers that spoke to the very points he raised. Horgan ignored them. That is dishonest."

Perhaps the most impassioned letter came from Berkeley's fiery anthropologist, Vincent Sarich, who thanked *Scientific American* for publishing "so concise a collection of muddled, misleading nonsense" to assist his "students in their eternal quest to inform and improve themselves through the mistakes of their elders." Sarich excoriated Horgan for "the obligatory connection of eugenics . . . and the Nazis." He then challenged the inference that knowledge of the gene-behavior connection leads to such excesses or that any one political view has a monopoly on atrocities. "After all," he writes, "Stalin and Mao, in the name of 'eumenics,' systematically murdered far more people than Hitler ever did. Human beings can do nasty things to one another appealing to almost any ideology, but surely the entirety of human evolutionary history tells us that knowledge is preferable to ignorance."

Bouchard continued to fight back at the Horgan broadside. Unsatisfied with being cosigner of a letter to the editors, he wrote a nine-page paper for circulation within the scientific community refuting only those parts of Horgan's piece that referred to his work but encouraging others whom he knew to be similarly misrepresented to do likewise. Most of the letter itemized the information Bouchard provided Horgan that put to rest his points and which Horgan did not mention, but Bouchard rarely got more accusatory than ". . . one must wonder why . . ." Measured as Bouchard's rebuttal was, his exasperation and anger surfaced at times, as when he confronted Horgan's claim that separated twins, excited by the attention their sameness brings them, strive to exaggerate their similarities.

Bouchard responded, "The twins would have to be pretty talented to exaggerate their similarities in I.Q., special mental abilities, reaction time to a variety of tasks they have never seen before, responses to thousands of questions they have never considered, as well as cardiovascular functioning, pulmonary functioning, cavities, etc. This is shoddy journalism with a vengeance." But of course Horgan gave no hint that such detailed measurements are made in Minnesota; instead, we are left to assume, some self-styled scientists sit around waiting for twins to do something weirdly similar, then call the newspapers.

Anyone familiar with the advances of behavioral genetics of recent years, and with the long-standing arguments against them, had to conclude that the article was shamelessly slanted and deliberately misleading. It confused and debased what could have been an honorable debate with false information and the omission of all data that might have rendered less "dubious" the link between genes and behavior. As disheartening as publication of the article was to the hundreds of men and women who have devoted their lives to this "dubious link," and to the large numbers of nonprofessionals who believe these scientists are on to a new dimension of human self-knowledge that could eventually be widely beneficial to everyone, all of these people were due for another rude shock. An organization called the National Association of Science Writers, which is based in Huntington, New York, voted Horgan's piece the best magazine article in 1993.

SEVENTEEN

VIOLENT CRIME IN THE MARYLAND WOODS

THE DRIVE TO THE Wye Conference Center on Maryland's Eastern Shore on a Friday evening in September 1995 placed me in an inching line of traffic heading to the beach resort Ocean City. The file of cars, city people from Baltimore and Washington eking another weekend from the waning summer, was so unrelated to the surrounding farmland, a sealed train from city to beach, that it added to the sense of rural isolation. I was not bound for the beach but on an errand to learn if behavioral genetics could be applied to one of the nation's most acute social problems, violent crime.

I turned off the highway and drove a few miles of country road, then found the center's private road for another mile or two of unspoiled landscape. Suddenly I came upon a car-filled parking lot behind which were a cluster of low, cedar-shingled buildings. After the miles of uninhabited landscape, the rows of cars came as a surprise, like stumbling on a Cosa Nostra meeting in the Adirondacks. This secluded and noiseless spot was the setting for the much debated, much decried, once canceled conference, now reborn with the title "The Meaning and Significance of Research on Genetics and Criminal Behavior."

The idea that criminals are "born that way" has probably been around since a Neanderthal, sitting with his tribe around a fire, first snatched a juicy hyena bone from one of his rule-abiding pals. The others surely gasped at this breach of caveman etiquette and concluded that the thief must be "different." What else could they think? The Neanderthals did not have today's bouquet of alternative explanations such as faulty rearing, lack of socialization, bad neighborhoods, peer-pressure, social stigmatization, racial oppression, and low self-esteem. Shaking their heads sadly, the Neanderthals had no choice but to assume a flaw in the thief's makeup. Of course, the bone-snatcher might have simply been hungrier than the rest. On such rudimentary contingencies endless environmental explanations can be built, many of them probably correct.

My Neanderthal parable makes a sad point about behavioral genetics: The further one descends on the scale of education and cultivation, the greater the acceptance of the idea that personality and character traits are innate, "in the blood." In our time, environmental explanations have been the stuff of learned journals and universities; hereditary ones the stuff of lunch counters and back porches. And to no group do the rocking-chair sages apply this diagnosis more frequently than to the most conspicuous carriers of "bad blood," criminals.

The effort to give scientific legitimacy to this cornerstone of folk wisdom began in 1870 with Cesare Lombroso, an Italian physician interested in the physical characteristics of lawbreakers. Lombroso believed he had stumbled upon an explanation of criminality when he was examining cadavers in the prisons of Pavia, hoping to find a pattern of *physical* differences between criminals and the insane. He came on the skull of a robber, a single skull, that sparked his insight about criminals.

Because of irregularities in the skull's shape, Lombroso hypothesized that many criminals were throwbacks to an earlier, morally inferior human. (It is interesting that he, like many, assumed that bad traits were older and more fundamental to human nature; the good traits, such as morality and altruism, came along as a part of our species' ineffable march toward civilization and virtue.) Lombroso developed his theory of "physical atavism" and published it in an influential book, *The Delinquent Man,* which emerged in 1873, a time when the entire concept of an *earlier* human was new.

Like so many spurious scientific ideas, Lombroso's fell on ears made receptive by visceral feelings, widely held then and now, that there must be *some* rational explanation for behavior that researchers, safe and untroubled in the middle class, considered perplexingly aberrant. In 1874, shortly after Lombroso found his link between physiognomy and criminality, Robert L. Dugdale traced the histories of 709 members of an Irish family, the Jukes, and found among them 76 criminals, 128 prostitutes, 200 living on public assistance, and 28 keeping brothels. While not denying an environmental role in the Jukeses' depravity (as did later researchers), Dugdale was convinced that the statistics on this one clan established a genetic basis to crime.

There would be many variations on Lombroso and Dugdale's ideas, each replacing the last and enjoying its fifteen minutes of acceptance in the scientific sunlight, but more important is that the search for congenital, as well as social, causes for crime had begun in earnest, giving birth to the discipline of criminology. Others prominent among those pursuing the inherited-learnings idea were Englishman Charles Goring and, in the United States, Herbert H. Goddard. The latter achieved notoriety for his study of the Kallikak family, whose generations of retards inspired the eugenics movement.

All the many variations on the genes-crime theme that have burst upon the scientific stage over the past hundred years have eventually slunk off, unproven or discredited, to a chorus of boos from the environmental crowd, those who had been rooting for failure. They invariably saw such theories as a ruse to block social-betterment efforts. They knew their opponents' stance: What point was there in trying to help the poor and disadvantaged who were driven to crime if they were irrevocably doomed to criminality by their genes?

The environmentalists had other good reasons to fear reckless applications of inheritance theories of crime. From roughly 1905 to 1930, the genetic foundations of criminality remained an important plank in the eugenicists' platform. Convinced that society's ills resulted not from economic inequalities but from innate flaws in troublemaking segments of the population, they unabashedly urged the sterilization of criminals—and succeeded. In 1915 thirteen U.S. states had enacted laws permitting sterilization of criminals and the retarded. (Virginia's notorious Lynchburg Colony was particularly enthusiastic in pursuing this policy.) By 1930 fifteen more states had done the same. These laws were challenged in 1927 in a case that reached the Supreme Court, *Buck* v. *Bell,* but the government's right to sterilize those it considered unfit was upheld in a decision supported by such libertarians as Louis Brandeis and Oliver Wendell Holmes (who wrote the famous line, "Three generations of imbeciles is enough"). Sterilization laws, while no longer implemented, astonishingly still exist.

This shameful episode in American history indicates that we do not have to cite Nazi atrocities to indicate the potential horror of hasty public policy based on wobbly science. As for the frequent allusions to the Nazis, Carl Degler points out in his book *In Search of Human Nature* that Germany joined other European nations in enacting sterilization laws in 1933, *before* Hitler came to power.

Even after the demise of eugenics as a legitimate approach to social problems, the search for inherited patterns in criminals continued, most notably with William Sheldon's theories in the 1950s about body types (watch your wallet around mezomorphs!) and the discovery of the double Y chromosome in 1965. Neither of these criminological breakthroughs stood up under scrutiny—a lot of very nice people turned out to have two Ys—but for a time, both laid claim to having discovered the biological underpinnings of criminal behavior.

With Lombroso's leap to a bogus conclusion, and with subsequent giddy leaps, all equally invalid, we see a mechanism that may help explain such sorry lapses in scientific judgment by otherwise intelligent men and women. Lambroso's perplexity over behavior he found unthinkable—good grief, *stealing*—toppled him over the edge of methodological soundness in his determination to explain how members of his own species could behave in ways so far outside his own behavioral range. Traces of such endearing naïveté can still be seen in today's social scientists who strain to unravel the mystery of behavior they consider "aberrant" only because its social context is so foreign to them.

As I ENTERED the Wye Center's main building and registered at a table set up in the lobby, I feared that I might be in for a more sophisticated variation of the same fuzzy thinking. A weekend on the Chesapeake discussing the reasons people commit crimes might reveal only that the question was as fatuous as asking Willie Sutton why he robbed banks. (Because my genes made me do it?) With or without valid conclusions, I had to acknowledge that the event's coming off at all said much about the idea's unsquelchability, given its history of misfires. For a group of distinguished scholars—scientists, historians, philosophers—to come together to consider the possibility of genetic links to crime and violence, and to do this in 1995, when the subject reeked of political incorrectness, was in itself remarkable. It was testimony not only to the idea's durability but also to the broadening acceptance of genes as an element in behavior-explaining equations.

The conference's coming to pass was even more remarkable because of an ill-fated effort by the organizer, David Wasserman, a legal scholar at the University of Maryland, to hold it three years earlier. The prior plans had progressed quite far, but a firestorm of controversy caused the event to be abruptly canceled. The title had been "Genetic Factors in Crime: Findings, Uses and Implications." With its assumption that crime *had* genetic factors, the title alone provoked angry accusations of a racist agenda and that the meeting was to be an opening ploy in a governmental plot to tranquilize the inner cities into manageable passivity.

Wasserman had crafted a proposal to allay the anticipated fears. As an example of his caution, he said that if genetic markers were found that placed children at high risk of becoming criminals, the markers would probably indicate little about who would and would not actually commit crimes. "Most people with the markers will not become criminals and most criminals will not have the markers."

As further evidence of evenhandedness and freedom from deterministic thinking, Wasserman and his colleagues had scheduled for their first meeting a review of the most recent failure in the search for a DNA-crime link, the YY chromosome. What subject could better demonstrate the dangers of premature conclusions? The conference's purpose, the prospectus said, was simply to examine recent behavioral genetics research on crime, critique it, and determine whether it had any public policy implications. When the proposal had been reviewed and approved by prominent academics in related fields, Wasserman received a National Institutes of Health grant of $78,000 and sent out invitations to all the leading scientists and sociobiologists on both sides of the issue.

Wasserman's diplomatic prospectus could not overcome disastrous timing. Sensitivities to the crime issue had been exacerbated a year earlier, when the Bush administration's Secretary of Health and Human Services, Louis Sullivan, had announced a "violence initiative." At a time when blacks were responsible for 45 percent of the nation's violent crime while comprising only 12 percent of the population, a move against violence was seen as a move against blacks. Warning lights flashed, and the politically watchful went on alert for the next manifestation of the government's racist intentions.

This was provided resoundingly by the Bush administration's director of the Alcohol, Drug Abuse, and Mental Health Administration, Frederick Goodwin, who in a speech cited recent monkey studies as having relevance to inner-city violence. Goodwin underlined his point by saying he understood why people called "certain areas of certain cities jungles." His remarks were the official toe out of the racist closet for which the vigilant had been waiting. Because of the resulting press furor, the violence initiative was abandoned and Goodwin was reassigned to a lesser post.

When only a few months later the Maryland violence conference was announced, many saw it, not as an attempt to bring new light to one of the nation's most critical problems, but as another sneak attack on beleaguered blacks. Leading the counterattack was a psychiatrist, Peter Breggin, known primarily as a dogged opponent of psychopharmacology, having fought every antidepressant from lithium to Prozac. Breggin joined a growing number who had learned a shortcut to prominence: making use of the media's admirable effort to grant equal space to both sides of controversies. As issues become less and less controversial, opponents become difficult to find. As a result the media must return again and again to the diehards, and some pretty marginal "experts" find themselves media stars. By being one of the nation's few psychiatrists who opposed pill therapies, Breggin was frequently in the papers. Similarly, he had little trouble snagging coverage of his outrage over Goodwin's monkey gaffe and was quickly the most vocal denouncer of the violence conference's sinister intentions. In Breggin's view, Wasserman's announced aim of a simple examination of genetic findings was a smoke screen for the government's hidden antiblack agenda.

Breggin's scare tactics succeeded. A Washington newspaper aimed at Afro-Americans typified the hysteria in a headline: PLOT TO SEDATE BLACK YOUTH. The Congressional Black Caucus was incensed to learn that government funds were underwriting bogus scholarship aimed at legitimatizing racist programs. The controversy became so inflamed that the N.I.H. had little choice but to withdraw the conference's grant. Wasserman's denials of Breggin's characterizations, plus angry counteraccusations of academic censorship, found little support in the academic community and the conference was canceled.

There was considerable irony in the Breggin camp's victory in that the conference would have included many powerful *opponents* to biological and genetic explanations of crime. So weighted was the guest list with militant antigenes types that the meetings might well have squelched, at least for a while, the notion of biological links to crime. By killing off the conference, Breggin had probably killed off a prestigious endorsement of his view that crime resulted from adverse social conditions, nothing more. In a conversation I had a year and a half later with Harvard's Jonathan Beckwith, a leading behavioral genetics skeptic, he expressed his regret that the conference had been canceled; he saw it as an opportunity to air the fallacies and pitfalls of efforts to link crime and genes.

In thinking over this drama, I felt that the conference had been precipitous. I didn't think there had been enough research on this one behavioral aspect. In addition, the entire subject of crime was so unwieldy, its terms hard to define. "Crime," "criminal," and "violence" are concepts that vary from period to period, from country to country, from neighborhood to neighborhood. Attempting a scientific analysis from the behavioral genetics perspective would be, to me, like shooting at not one moving target but several, all passing in different directions, with one bullet, genetic theory, to hit them all.

I also saw Wasserman's project as typifying a lamentable pattern in the history of genetic approaches to human nature. Each time advances have been made in understanding of the biological links with behavior, ideologues with extreme positions appear, cheer the new findings, then twist them to their own political agendas. As a result they stigmatize the field and succeed in nothing more than driving the line of investigation back into the no-man's-land of intellectual disrepute. A crime and violence conference at this point in the unfolding science of behavioral genetics might appear another replay of this counterproductive scenario.

MANY IN THE FIELD disagreed. Shortly after the collapse of the Maryland conference, I was walking across a snowy University of Minnesota campus with Thomas Bouchard and David Lykken, on the way to lunch. I asked what they thought about the conference controversy, hoping that they would share my opinion that it had been premature. They didn't. Without answering me, Bouchard turned to Lykken and said, "You know, David, it's time for that conference. We should think about holding it here." Lykken nodded a thoughtful assent.

I knew that neither Bouchard or Lykken harbored the sinister political motives that Breggin and the other critics imputed to the Maryland organizers. Behind the two men's desire for a crime conference, I suspected, was an eagerness to link their behavioral genetics discoveries to issues of broad national concern. A high-visibility conference would be a way to establish with the public the relevance of their science to the nation's life. But they could not expunge the political taint to the subject, and I did not believe the data were yet strong enough to guarantee safe passage through a p.c. minefield.

Yet I was a newcomer to behavioral genetics, and I was talking to two of its most respected and knowledgeable practitioners as well as two veterans of the wars between science and ideology. As I listened to them discuss a revived crime conference, I felt like a hound dog watching helplessly as two admired humans waded into quicksand. The most I could muster as a warning yelp was to ask Bouchard if he thought a crime conference was wise in light of Maryland's collapse.

He looked at me blankly. "Why shouldn't we hold it? We don't get any government money."

I was stunned. Bouchard was surely aware of the press attacks on the Maryland conference, the *ad hominen* invective from the Peter Breggin types, the denunciations from the black caucus in Congress. Yet he was brushing this aside to focus on only one factor: funding. He was either very brave or very reckless. Or better informed than I. After a few hours on the university's database, I learned that there was more research than I realized on the genes-crime connection. Apart from animal breeders who for centuries had been breeding hyper-aggressiveness into bulls, attack dogs, and fighting cocks, neurologists and clinical psychologists had long known about biological links with various forms of rage and violence.

In experiments headed by M. Glusman, a small section of the hypothalamus of cats was removed, rendering the animals perma-nently savage. Certain humans who had suffered head injuries and brain insults would be plagued for the remainder of their lives with uncontrollable fits of violent rage. The brain-chemistry-violence nexus had been well demonstrated in prisons by the effectiveness of lithium and beta blockers in reducing violent aggression in trouble-some inmates. Experiments with violent children had shown similar pharmaceutical interventions to be highly effective. In sum, the strong links between brain chemistry and violence had for years been a commonplace among psychiatrists and neurologists, even though soci-ologists were having none of it. The genetic contribution to criminal-ity was less well established.

But even here evidence existed. A meta-analysis of twin studies based on fraternal-identical comparisons showed that in eight studies of aggression, identical twins were concordant on an average of 49 per-cent of the time, versus 29 percent for fraternals. On measures of im-pulsiveness, thirty-one studies averaged 51 percent concordance for MZs, 20 percent for DZs. The Minnesota reared-apart twin study also found a significant genetic component to aggression and "self-control." Aggressiveness and impulsiveness were clearly influenced by genes, but while such broad traits might be *conducive* to violent crime, their presence, whether from genes or the environment, was hardly, well, incriminating.

The most impressive study actually linking crime to genetic makeup was conducted in 1983 by Sarnoff Mednick, who organized a team to examine 14,427 convicted Danish criminals. All of those stud-ied had been adopted between 1927 and 1947. Because of the exhaus-tive records kept by the Danes, Mednick was able to compare the criminal histories of his sample with those of their biological parents. The study established a clear genetic involvement with property crimes like burglary and embezzlement, but *none* for violent crimes.

This was surprising because of the more pronounced, definable nature of violence and because of the accumulated data on the biological mechanisms that produce it. Widely reported findings on the relationship between violence and the levels of the neurotransmitter serotonin had weakened scientific resistance to the notion of biochemical links to hyperaggressiveness. Violence appeared to be a trait ripe enough to drop in the behavioral geneticists' laps. Why Mednick's study failed to establish a gene-violence connection and so many other studies have done so is an enigma that no one, as far as I know, has explained. Perhaps science would get boring if the evidence always pointed in the same direction.

FRANK ELLIOTT IS a retired professor of neurology at the University of Pennsylvania. Originally from South Africa, Elliott is a soft-spoken man, now in his eighties, who has spent much of his career observing and treating brain disorders in the criminally violent. Like many familiar with the physiological contributions to violent crime, Elliott has watched in dismay as criminologists examined only the social and psychological causes. Years of clinical experience had convinced him that two forces were involved—one environmental, one physiological—both with much to tell the other. Most troublesome to him was the way one of these camps, the social sciences, resolutely ignored the other. When I met with Elliott, he told me of his great satisfaction in having witnessed, in the past few years, a gradual acceptance among criminologists of biological and genetic influences on at least some manifestations of criminal violence.

In a 1990 paper Elliott wrote: "Clinical experience with violent individuals of all social levels suggests that it might be more useful if such violence were to be regarded not solely as a product of psychological disturbances or social adversity or physiological deviance, but as a net behavioral result from the confluence and interaction at any given moment of multiple biological and environmental variables, some excitatory and some inhibitory."

Reinforcing the scientific data and pointing doggedly to a some-are-born-that-way approach to criminality was the statistic about adolescent male violence that Frederick Goodwin so often cited: 79 percent of repeated violent offenses were committed by 7 percent of the youth. Seen from another angle, about one-eighth of the population commits about one-half of the crime. These and similarly evocative statistics hovered over all of the discussions, and would hover as well over the violence conference, to serve as pesky reminders that for all the talk about poverty, racism, oppression, and hopelessness, within the segment of the population exposed to these negative forces, *only a small percentage* suffering these social ills turned to violent crime. Nonenvironmental causes also appeared to be involved.

Bouchard and Lykken's hopes for a reborn violence conference were shared, it turned out, by the beleaguered David Wasserman. Despite the pummeling he received in the press, he was convinced of his project's timeliness and that *not* holding the conference could be seen as a tacit admission of the conference's underlying racist motives. (You found us out. The game's over.) Because his plans for a resurrected conference were well advanced, he spared Lykken and Bouchard the trouble of staging their own.

That the conference was finally coming off was a tribute not only to Wasserman's determination but also to the spreading acceptance of a genes-behavior connection within the scientific community during the two years since the prior cancellation. Increasing numbers of socially concerned people believed that society's most acute behavioral problems—like crime—might be better understood if the new genetic information were factored in with known environmental causes—poverty and racism, in the case of crime. As I found a seat in the lecture hall at the Aspen Institute's Wye Center, I thought that the next two days would tell if the idea were premature or not.

THE KEYNOTE SPEAKER was Irving Gottesman. Because of his reputation as a founder of behavioral genetics, this choice in itself suggested a hospitable attitude toward genes-crime theories. In fact, the list of those in attendance showed that the participants were not preponderantly opponents, as in the first conference. Also telling, the best-known critics of behavioral genetics were absent. I later learned that both Lewontin and Breggin had accepted, then canceled, perhaps when they sensed a tilt from their antigenes view. Beckwith was not present, nor was Kamin or Hubbard. The only two names I recognized of the Cambridge antibehavioral genetics group were Diane Paul and

Evan Balaban, the latter having established his hostility to the conference's subject by telling Natalie Angier of the *New York Times* a few days earlier that he had "a strong opinion that biology doesn't have anything to contribute to public policy discussions about crime in society."

As advocates for their side, they were outranked by some behavioral genetics stars. In addition to Gottesman, we would be hearing from Gregory Carey of the University of Colorado; Adrian Raine of the University of Southern California; and Martin Daly, the co-author of *Homicide*, the landmark book on the evolutionary underpinnings of murder. There were several neurologists and psychologists who worked with problem children and two black academics, a fair representation although Wasserman had tried hard to lure more into participating.

In his talk, Gottesman stressed that genetic effects are usually intricate interactions with environmental factors and that for complex behaviors like crime, genetic involvement, when present, is even more complex than for other traits, probably involving many genes. He strongly disavowed the notion of "a crime gene," calling the phrase nonsense. A genetic link to crime might be nothing more, he said, than a neucleotide predisposition in some people to behavioral tendencies such as impulsiveness that under certain conditions make the probability of criminality higher than it is in most people. He also said that two prisoners may be in jail for the same crime, but one was impelled by schizophrenia while the other was reacting to an oppressive environment.

Gottesman pointed out that some violence results from genes common to all members of the species, implying that since we are not all violent all of the time, in those instances the inducements had to be entirely environmental (retaliation, self-defense, and others). He stressed, too, that genes are dynamic, not static, that they turn on and turn off continually throughout life. With all his disclaimers, however, Gottesman emphasized that in trying to understand causes of crime, genetic factors must be addressed. He cited reared-apart twin studies that showed a genetically significant concordance in criminal behavior, noting specific instances of highly similar arrest records between twins.

After dinner Friday we listened to three speakers, two of whom spent most of their time proclaiming their sensitivities to the dangers of applying genetic information to crime. They all saw crime as mostly a social issue, not biological. The only one who put in a forceful "Yes, but . . ." was a criminologist from the Department of Justice, Diana Fishbein, who began her talk by stressing that her opinions on the conference's topic were her own and in no way reflected official Justice policy.

Fishbein said that in twenty years of attending criminology conferences, she heard the same people air the same ideas with no reference to possible physiological causes. In spite of compelling data in recent years, biology was still omitted from the criminologists' equations. These crime experts, she said, remained resolutely uncurious about the reasons all people exposed to the same environmental circumstances don't react in the same way. They left this inconvenient question to the psychologists and others who were engaged in behavioral genetics research. Although criminologists were the best positioned to pursue the biological line of inquiry, they stubbornly resisted even considering the data others were producing. She felt that this was a serious mistake.

Her main point was that in spite of the dangers, enough was now known of the biochemical basis of some types of criminality for pharmaceutical interventions to be considered as public policy in high-risk cases. Fishbein didn't stop there. In dealing with repeated perpetrators of violence, she believed drug therapy should be *mandatory*. She was talking about mild stabilizers like lithium and Prozac. Although the latter was then being consumed voluntarily by six million Americans, to many of the audience, such talk was the first step in the government's suspected agenda to drug inner-city youth into a manageable passivity.

I later spoke with Fishbein and complimented her on being the only one of the evening's speakers who came out and spoke candidly about the auspicious potential of the new behavioral genetics information. The few similar remarks made by others had been lost in a morass of caution and social awareness. Fishbein was grateful and told me that she had doubted if participating in the conference would be worth the effort, so certain was she that her recommendations would meet strong resistance. Her obvious expertise and her conviction won the respect, if not the agreement, of the mostly skeptical audience.

Violent Crime in the Maryland Woods

.   .   .

OVER THE NEXT TWO DAYS, I was impressed by the scientific acumen and eminence of the participants. I was also impressed by the high level of civility among people with diametrically opposing views. Counterattacks were couched in terms like "I think you overlook . . ." or "perhaps you are unaware . . ." Mercifully, there were no slurs about racists and Nazis; I don't even recall a single "naive" being lobbed across a panel.

Defining fundamental terms proved a nagging problem. Even on the last day, speakers were still arguing over the meaning of "heritability." About half of the speakers pointed out the vagueness of concepts like "crime" and "criminality," saying they were vague social constructs not conducive to scientific study. Frank Zimring, of the University of California Law School, amused the crowd by saying that sodomy is a crime in many places but that large segments of the population "think it's terrific." Gambling was another example Zimring offered of a crime enjoying widespread popularity; he cited heresy as a crime that had had its day. Other speakers were persuasive in establishing that for an inner-city male youth, criminal behavior was not pathological but rather normal reaction to dire circumstances.

One speaker said that the turf wars of South Central Los Angeles were no different, in terms of evolved human behavior, than territorial disputes between nations such as the Gulf War. This was one of those generalizations that cried out for dismissal as too broad. But a quick brain scan for the illogicality or incongruence, with me at least, turned up no handy rebuttal, and the parallel stood, serene and radiant, as a one-sentence denigration of the entire concept of war.

On the progenes side, a few speakers made the point that until we understand the genetic factors involved in criminality, we cannot pinpoint the most important environmental factors. Another argument that seemed to slow the critical momentum was that when high-risk individuals in high-risk environments are found, scarce government resources for recreation, job training, drug therapy, and counseling can be targeted on these people rather than spread throughout a large population, most of whom are in no danger of becoming criminals. (While reasonable, this idea raised the cloud of a different brand of injustice: the bad guys receiving all the governmental goodies.)

While the proponents of seeking out biological influences on criminality made clear that the prime beneficiary would be society, they also emphasized that the criminals themselves would benefit as well. Speakers who worked with delinquent youth said that many of the most problematic kids were eager to change. One biologist even addressed the worst-case bugaboo, government sedating inner-city youth, and asked, "Isn't sedation better than twenty years in Attica?" The biology-oriented were emerging cautiously from the closet.

(This hobgoblin of government eagerness to decriminalize trouble-makers medically was thrown into interesting light in April 1996, when a Texas child molester, Larry Don McQuay, who was about to be paroled, pleaded with the state to castrate him, saying that when a free man, he would surely kill more children. Horrified, the state refused.)

Noteworthy among the speakers was David Comings, a geneticist at the City of Hope Medical Center near Los Angeles, who works with troubled young people, many of whom suffered from attention deficit hyperactive disorder. Genetic research had shown that the genetic anomaly that causes ADHD is closely linked with other problems, such as alcoholism, drug abuse, and criminal behavior. Comings told the group that he had treated "thousands of these kids . . . who are failing in school, unable to concentrate. Teachers find them disruptive, aggressive, fighting, lying. . . ." Their families had brought them to Comings in desperation. Various pharmaceutical treatments had totally altered the unwanted behavior. Comings surely confused the drug-fearing liberals in the audience by adding that these treatments are expensive and at the present only available to white middle-class kids.

In her talk, Diana Fishbein had said that as she saw it, the hope was to pinpoint crime- or violence-related genes, in order to identify and aid those most likely to fall victim to adverse social conditions. Throughout the remainder of the conference, the central argument seemed to devolve on her one word *aid*. Most participants agreed that identifying those at risk for asocial behavior had been made feasible by the new genetic knowledge. There was sharp disagreement, however, about whether this knowledge would be used to *aid* potential criminals or would instead be used to stigmatize and isolate them, to inflict further injustice on the disadvantaged. Over the next two days the speakers tended to align themselves with one of the two positions.

Another theme sounded frequently by the opponents of pharmaceutical crime fighting was that the quest for biological links to violence showed a yearning for "simple solutions," the old idea that if everyone's destiny is gene determined, social improvements are pointless. As often happened, these critics appeared to be arguing with turn-of-the-century eugenicists, not with the behavioral geneticists present, all of whom emphasized the complexity of the genes-environment interaction and the importance of poverty, racism, and the rest in causing crime.

Toward the end of the second morning's session, I was beginning to marvel at the spectacle of avowed opponents of behavioral genetics applications to social problems like Evan Balaban sitting attentively while various scientists talked about hormonal and genetic anomalies conducive to crime. Perhaps the skeptics had been soothed by the prolonged lip service every speaker was paying to the importance of environmental factors, conditions that were society's fault and society's job to correct. Or perhaps they were waiting for their moment to pounce with the usual *ad hominem* denunciations.

IF IT WAS THE LATTER, they were spared the trouble. Lunch was only one speaker away when I heard shouts coming from a terraced area outside the rear of the meeting hall. It sounded like children in a playground, but I dismissed that as unlikely at a conference center aimed at quiet adult ponderings. The shouts grew louder. Suddenly the doors burst open and a stream of protesters pushed their way into the hall—young blacks and whites, middle-aged white men and women, about thirty-five in all—and started down the two aisles. Some carried placards; others brandished red flags. Some had bullhorns and all were shouting a chantlike refrain that at first I couldn't make out but which emerged as a reference to the meeting's rural isolation:

"Genetics conference you can't hide, you're promoting genocide."

As the group moved raucously down the aisles toward the stage, it seemed that we were about to experience firsthand a sample of the behavioral trait we were there to discuss. I later learned that sympathetic members of the audience had been in contact with the protesting group and had agreed to open the locked fire doors at a pre-arranged time. One of the band shouted they were here to shut down the conference. Others voices emerged with cries of "racists" and "Nazis."

As they reached the podium, the scheduled speaker docilely yielded the microphone and stepped aside. With this ultimate victory, the protesters arranged themselves in a semicircle on the floor below the stage, a maneuver that placed them inches from the first row, where I was sitting in order to tape the talks. Some of the group proclaimed themselves members of the Communist Party, whose existence in rural Maryland came as a surprise to many in the audience ("In Maryland? *Anywhere!*" said one professor). As one speaker stepped aside to give the microphone to a cohort, the protesting chorus filled the lull with the chant "Fight for Communism! Power to the workers!" and later "Jobs, yes! Racism, no!"

Encountering no resistance to their takeover of the amplification system, the intruders lowered their belligerence, but an ugly mood of coercion roiled the auditorium. We would listen to them whether we wanted to or not. No one in the audience moved. Whether from fear, reluctance to miss the spectacle, or both, we were fixed in our seats. A succession of speakers, taking turns at the lectern, hammered at the theme that inner-city crime flourished because of poverty and hopelessness, a lack of alternatives—no other reason. By searching for genetic reasons we were attempting to shift blame from oppressive conditions and place it on oppression's victims. The reiteration of themes we had been hearing for many hours in the scheduled sessions pointed up the protesters' obliviousness to the conference's range of opinion. A distraught Wasserman tried to yell this at those controlling the stage but was shouted down.

With a line of protesters directly over me, I resolutely looked beyond them to the stage speakers. With my face a study in rapt attention, I felt sure anyone would see the thoughtful consideration I was giving the rhetoric spewing from the loud-speakers as the demeanor of one about to be converted from nasty genetic determinism to benevolent social awareness. Since I was not of the university world, I was unsure my fear was appropriate. I assumed the academics around me were accustomed to such disruptions and were unperturbed by the chest-thumping bellicosity. I was unnerved when a male molecular biologist, husky and tenured, leaned toward me and whispered, "I'm scared."

From the microphone, a middle-aged woman in a babushka told us that even though our intentions might be well meaning, the genetic information we sought could be used by others for evil ends, as it had been in the past. Like many of their points, this was a valid and now-familiar admonition but with the unavoidable antiknowledge thrust. Still, I felt relief to have our characterization by the invaders change from genocidal racists to unwitting dupes. You don't hurt unwitting dupes, I reassured myself; you enlighten them. With the new tone I felt the tension lift.

A young black male, clearly irate at the conciliatory drift, tried to recapture the combative mood by shouting that if we attempted to enter the inner cities with our sedating drugs, the people "would be ready and will initiate some violence on y'all." Then for a punctuating punch, he added, "You want trouble. We'll give you trouble." His companions did not build on his bellicosity. No one was rude enough to point out that for the government to bring drugs to the inner cities might be seen as redundant.

A man in the audience succeeded in winning the protesters' attention. "You don't want to talk with us," he said, "we are small beer. You want to talk with the media and the public." With exaggerated magnanimity he added, "Let us make you a gift of all the media here today."

A derisive laugh from the seated academics signaled a realignment of adversaries I found worrying. Until now it had been the protesters versus the conference, but the remark pitted both groups against a handful of journalists—of whom I was one. It was upsetting enough to have an angry band looming over the audience calling us racists and fascists. Now the conference participants were themselves turning on my small subgroup. I looked around for a fellow journalist.

An offer was made to the occupation forces of a formal press conference on the terrace outside the auditorium, probably a ploy to ease them out of the building. Fortunately, they agreed, with flags lowered and sullen faces, they filed up the aisles. Reassembled outdoors, they repeated the same slogans and rhetoric (and revealed that they had organized their demonstration on the Internet) but now to a smaller, but more like-minded, audience. The majority had gone to lunch.

THE REMAINDER OF the conference was tranquil and informative. If the disruption had any effect, it was to make the panelists even more courteous to each other. The content of the sessions remained short on science, as some had predicted, and long on social awareness. It had also ignored one of the most chilling aspects of much of today's violent crime: black hatred of whites, although that may have been hidden under the euphemism of "social conditions."

For all its digressions, disclaimers, and disruptions, the Wye conference had succeeded resoundingly in its primary aim. It had opened a dialogue about possible applications of genetic and biological knowledge to the problems of crime and violence, applications that could be compatible with a vision of a just society. The conference, and the startling new knowledge it examined, had not been shut down by the protesters, nor had its science been crippled by antigenes, highly political academics who saw it as just another brushfire in the retrograde move toward genetic determinism.

The distinction usually made between the twin disciplines of behavioral genetics and evolutionary psychology is that the former is concerned with the genes-based differences between individuals and the latter focuses on traits common to all members of the species. The 1995 violence conference turned out to be a three-day effort to decide to which of the two disciplines crime belonged. Was it a trait that found expression only in certain individuals? Or was it a species-typical behavior that would emerge in all humans under the right—or rather, the wrong—conditions?

By the end of the meetings, the conclusion seemed to be it was both. Yes, under certain circumstances everyone is capable of violence—and the plight of inner-city black male teenagers, most agreed, would qualify as one of those circumstances. And yes, certain individuals, because of their genetic makeup, are far more prone to violent crimes than others. Ancillary to this thought is that with the violence-disposed, a pill regimen could stop the slide into trouble. Among those who had repeatedly been criminally violent, many could be medicated away from this behavior. Whether such intervention would be the criminals' choice or society's was a value question, not a scientific one, and a question left almost untouched by the discussions.

I saw an irony in this conundrum of the two types of criminals— one the result of social conditions, the other the result of flawed neuro- biology. While the conference members, heavy with degrees and expertise, grappled with recognizing two categories of criminals, it amused me to think that gang members in South Central L.A. would have no problem differentiating between the two. They undoubtedly knew that among their buddies, some were "different" in the hard- wiring department. "Sure, we mug and rip off stores," I could hear them saying, "but Pablo is *crazy*. He'll kill you as soon as look at you."

With learned professors in Maryland pondering whether the cause of violent crime is cultural or genetic, gang members immured in crime know that for some it's one, in the sense of what's going down, and for some it's the other, in the sense that they are "different." And they know which is which. For social planners the question remains: Can and should Pablo be identified early on and eased away from his probable path—by counseling, job training, perhaps medication? Or should he be allowed to commit X-number of murders before society considers pacifying him chemically? Or neither? At the moment, soci- ety's only response is prison or execution. The Maryland conference was attempting to find alternatives. As for the equally large question of why there was so much more violent crime in the United States than elsewhere, the conference offered no answers.

ABOUT THE TIME of the Maryland violence conference in the fall of 1995, a book called *All God's Children* by Fox Butterfield attempted to answer this question. It was the story of a black family, the Boskets, who had a standout history of violent crime. The author, a respected correspondent of the *New York Times*, traced the violent traditions of the South Carolina county from which the Boskets came, the Anglo-Irish codes that triggered murderous responses to slights to honor, the absorption of these mores into black culture, and their eventual importation to big-city slums. (That this same justification for fights to the death seems also to have traveled from South Carolina to every corner of the globe is not addressed.) The book also examines the "bad" black man as folk hero, the elevation of mavericks and troublemakers to legendary status in black culture. Always present in the narrative is the mood of hopelessness and futility such heroes helped assuage. Having established this historical framework, Butterfield relates the horrendous childhoods of the Boskets, their repeated lawbreaking, and their prison careers, which began before puberty and had few interruptions throughout their lives. In all of this the author sees "not just a portrait of the Boskets, but a new account of the growth of violence in the United States."

In a day when enlightened people are turning to genes and biology for help in explaining every form of behavior, normal as well as aberrant, *All God's Children* serves as a monument to the tenacious belief that the important answers lie in the environment if we just sift through it carefully enough. As a reporter, Butterfield is too thorough not to allude to possible genetic contributions to the Boskets' extreme behavior, but it is the merest lip service, polite nods to current trends of thought—a few paragraphs in 331 pages of environmental exposition.

Most of the book relates the sorry family's saga with little reference to other blacks of their world. Butch did this; Willie was sent there. On the few occasions when an incident throws them into peer-group context, like Willie's unparalleled mayhem at the reform school Wiltwyck, we see him as a stand-alone troublemaker, a rebel among rebels, and an altogether atypical case history representative of little, I would guess, except his own exotic neurobiology. In Leavenworth or any of the prisons to which the Boskets were sent, the family were unique pariahs whose extraordinary viciousness won them the fear and awe of the violent criminals surrounding them.

The supposition of Butterfield's book is that the Boskets were, like all violent blacks, products of the same southern traditions mixed with oppressive and abusive rearings, as were their fellow inmates. Their violence stemmed from the same cultural and social causes but simply took more intense forms for reasons left unexplained. The implied premise is: You've seen one violent black family, you've seen them all. In spite of the genetically evocative factor that they are a *family*—not a gang, not childhood chums, not prison mates—this biological link does not trigger curiosity about the possibility of a genetic contribution to their extreme behavior.

The Boskets' violence makes for interesting reading, to be sure, but their bizarre stories tell less about the social and cultural forces of crime than would the stories of more routine criminals. We are left in the dark about the factors that set them apart from other violent men and women. And we are left to wonder if all the exacerbating elements of their environments, so meticulously documented, may have been nothing more than convenient pretexts for violence the Boskets would have committed regardless of their background. Comfortable middle-class homes have produced ax murderers and serial killers, but in such cases, if we search hard enough, and we usually do, a cold mother or an overbearing father can usually be found to serve as environmental culprit when we can't fall back on racism and poverty for explanations. In today's preoccupation with social injustice (probably a lingering obsession from the Marxist and Socialist heyday when it explained everything except meteor showers), such household-specific idiosyncrasies might be viewed as the second team in the pretext catalog.

The grim facts of the Bosket histories undoubtedly played a role in their chronic explosions. These background elements surely reinforced, even enabled, behavioral bents whose primary causes may have resided in the family neural structures. But we are given no information to allow us to sort out the environmental from the genetic causes. For instance, we do not learn the proportion of violent black males produced by this one South Carolina county, a crucible of violence in Butterfield's characterization, compared with violent black males from other, less violent counties, or for that matter compared with the entire United States. Even more important, we do not hear about the hundreds and thousands of young black males who came from the same part of the country at the same and with the same dire childhoods who never turned to violence and crime.

By devoting close to three hundred pages to the Boskets' environments, most of which was prison, the author, clearly trying for an even-handed analysis, is ipso facto buying into Willie Bosket's self-serving blather about being a monster the system has created. We know that Willie is a product of his environment mainly because Willie says he is. And of course, Willie has had impressive substantiation of that theory by mainstream criminological thought of the past fifty years.

Butterfield's depiction of the cultural forces in which the Boskets developed surely tells a good bit about the origins of violent crime in the United States. Because this was his announced purpose, it may seem a quibble to put forward the opinion that the Boskets themselves do not. Although this is not said in the book, they typify a dilemma that law enforcement people and social planners will be facing more and more: to determine if a physiological anomaly has contributed to violent acts and if this anomaly can and should be mitigated pharmaceutically, or perhaps even with genetic manipulations.

The book, on the other hand, implicitly proffers the same strategy that has prevailed throughout much of this century: seek out and correct the relevant environmental problems. Not only are they assumed to be the principal causes, they are *alterable*; the genetic ones are not. Neither part of this pat analysis is still true. It would be foolish indeed for the government that presides over the highest murder rate in the world to ignore the scientific insights that offer hope in reducing the grim statistics.

In a conversation with a historian, I asked why we always looked to history for help with the world's bafflements and problems. Without hesitation, she replied, "It's all we've got." In terms of human behavior, I believe this is no longer true. The historical context is invariably relevant, with nations as well as individuals, but we can no longer feel confident that the past tells the whole story. We now have other places to look for answers. With an extraordinary family like the Boskets, their history may not reveal the cause of their violence but may merely be a symptom of the real cause, their singular biochemistry that is passed down generation after generation in their genes.

More recently, a similar tunnel vision marked a book about the sexual abuse of a retarded girl by a group of affluent high school boys in a New Jersey suburb. What aspects of our culture, what dark side of the American dream, the author asks, does this incident reveal?

While no one would argue that American cultural values might play a role in this incident, to seek the answer only in such environmental factors is ignoring what must be, at least, a major part of the precipitating equation. It might be more astute to reverse the question and ask what positive cultural values, the kind that restrain and forbid such behavior, failed to function in preventing impulses that may well exist in every adolescent male on the planet? With our new knowledge about biochemical elements, it is time to pull the camera back, when giving book-length consideration to alarming and puzzling incidents, and view them in the entire context of human makeup. The social values of Glen Ridge, New Jersey, and flaws in the "American dream" are no longer sufficient.

SHORTLY AFTER THE PUBLICATION of *All God's Children,* a research paper about mice was published in *Nature* that may well have said more about the Boskets than the southern tradition of violence and the brutalities of the nation's prison system. Geneticists discovered that by shutting off a gene responsible for producing nitric oxide, which acts in mice and humans as a neurotransmitter, the mice became murderous. Rigid rules of mouse behavior were viciously broken by the chemically deprived mice. Fights ended with a kill rather than the typical submission of the loser. Female mice not in the mood for love can ordinarily repel with little effort an amorous male; they were now raped mercilessly by the nitric oxide deprived.

This chemical link to behavior took on even greater interest four months later, when the *New York Times* reported a major discovery: that human hemoglobin, in addition to transporting oxygen to all parts of the body, also transports, of all things, nitric oxide. This was a particularly surprising finding, since hemoglobin had been meticulously studied since the 1930s and was thought to be fully understood. But for all those years no one knew that nitric oxide had been invisibly piggybacking on it.

Early in 1996, scientists working through Duke University discovered that the gas, as a bloodstream stowaway, intrudes on every aspect of human health, including blood pressure, memory, even erections. The *Times* made no mention of the earlier discovery of the link between nitric-oxide deprivation and violent behavior. Even so, the two discoveries—its ubiquitous presence in the human body and its connection to violence and other behaviors—strongly indicated that it might replace serotonin as the neurotransmitter of the year. It should certainly be the target of vigorous new research.

When nitric oxide's violence connection was announced in the

New York Times at the end of 1995, the usual naysayers were granted their usual space to warn against extrapolating from mice to humans—although the neurobiology of mice and humans has repeatedly been shown to be highly similar. Adding to the gratuitousness of the warnings was that none of the geneticists involved in the study claimed anything more than that the findings suggested *possible* implications for human behavior. At a time when rape and murder were common occurrences in the United States and domestic violence loomed as routine, I would have thought the discovery that a neurotransmitter has a direct bearing on violence—in mice, baboons, or fruit flies—would have been greeted with the keenest interest, not with the automatic alarms of those who dread biochemical explanations of behavior and fight off any information that narrows the gap between humans and other creatures.

The months following the Maryland conference were a particularly rich period for crime-related research findings. In February 1996 yet another study was released that had important ramifications for the unfolding understanding of the chemical mechanisms underlying behavior. A research project at the University of Pittsburgh proved conclusively what many had suspected for years: A connection existed between high blood levels of lead and hyperactive, antisocial, even criminal, behavior. The main ways lead gets into humans are from flaking pre-1980 paint and antiquated plumbing, both commonplaces of the inner cities. Thus, the link between lead and crime had powerful societal implications. A study group of children, traced from ages seven to eleven, demonstrated that those with high lead levels were far more likely to become troublemakers and criminals.

The lack of nitric oxide that turns mice into killers results from an anomaly in their own genes; the lead contamination comes from the environment. Both problems—one caused endogenously, the other exogenously—are only two of the most recent indications of links between chemical imbalances and criminal behavior. When the back-porch biddies spoke of criminals' "bad blood" they appear to have been more accurate than they—or any of us—knew.

These chemical links to violent behavior, however, by no means imply that adverse social conditions aren't still an important, perhaps the most important, cause of crime. Still, if pharmaceutical or other forms of help can be given to those whose dangerous criminality stems from their internal chemistry, even if these individuals represent only a small percentage of the criminal population, our society would be foolhardy indeed to ignore such research because of paranoid fears, so well articulated by the Maryland protesters, of sweeping governmental plots to sedate black youth. We should harness the new scientific knowledge to reduce violence in this country, and perhaps enlist the genetic watchdogs to make sure the capability is not misused.

EIGHTEEN

CONCLUSIONS

BEHAVIORAL GENETICISTS JOKE that people with one child are environmentalists; those with more than one are geneticists. The story implies a self-evident aspect to the inheritance of personality: that children demonstrate temperament and behavioral differences at too young an age to allow environmental explanations. But if all this has always been obvious to parents, it was not just ignored by science for the past fifty years, it was vigorously denied—and still is in some Foucault-besotted corners of academia.

Kenneth Kendler theorizes why many people, including scientists, resist accepting a genetic influence over their actions—for other than the usual political reasons. "They fear," he said, "that would mean they are not masters of their own ship." As I hope has been demonstrated in the preceding pages, no such glum conclusion is necessary. The new information asks only that we recognize that the vessel we presume to control has predispositions—has, in fact, a mind of its own. Fortunately, we do too. The fiction of despotic genes usurping our sovereignty over ourselves has by now, I hope, been laid to rest. As far as behavior is concerned, there is no genetic *determination*, only genetic influence.

Whatever the reasons for resistance to the genes-behavior connection, the preponderance of scientific evidence is at last coming into synch with the evidence of most people's experience, so that gut hunches are now reinforced by hard data. Yet, the persistence of the environmental thought patterns is discouraging. While genetic explanations are becoming acceptable for temperaments and talents, the tendency toward rigid environmentalism springs back to life when the dialogue shifts to more complex behavioral manifestations—depression and rebelliousness, for instance, and other traits traditionally thought to be products of dynamic mental process. Even now, when these longtime domains of traditional psychology are discussed, the usual suspects are trotted out—rearing, home setting, education, role models, and the rest—with little more than lip service to possible biological contributions—and this is true among educated people and in some alert, up-to-speed journals. The environment-scanning habit goes very deep.

With the critics' unceasing cries of alarm about the perils of a genes-based view and the impugning of the motives of those who hold it,* behavioral geneticists are reduced to denying hidden agendas or to demonstrating—as I have tried to do in this book—that the implications are not as dire as the gene police would have us believe. Often lost in the argument are the many benefits that will come from the new understanding of human personality and behavior.

Perhaps the most auspicious benefit is the possibility of interventions and remedies for unwanted psychological conditions that till now have been considered as intractable or reachable only by extended therapy. The pharmaceutical adjustments to chronic chemical imbalances—Prozac, Zoloft, Paxil—have been amazingly successful. While these drugs can be seen as a massive rebellion against genetic influence, they merely adjust neurological abnormalities in which genes often appear to play a part. With the rapid advances in molecular biological knowledge and such landmark research as the Human Genome Project, the direct interventions that have already been successful with physical problems may soon be possible with behavioral problems as well. Scientists are already predicting that as the genetic basis for various forms of addiction, depression, and sexual aberration are pin-

---

*The April 21, 1997, issue of *U.S. News and World Report* quoted Leon Kamin as saying that the simplest way to discover someone's political leanings is to ask his or her view on genetics. Old myths die hard.

pointed and understood, nucleic adjustments might one day become available to whomever wants them.

Potential misuses are, of course, a danger, as was stressed at the Maryland violence conference. Governments could emerge that forced "cures" on its people; this might begin with such clear-cut problems as criminality, but tyrants might push ahead to such "problems" as an unwillingness to be oppressed. They might well see this attitude as a "condition" that needed remedying (as the Russians saw it as mental illness needing hospitalization). Happily, in this country at least, the government has shown a marked reluctance to employ drug therapies that could reduce the pandemic problem of crime. There is less reason to think a future government would be eager to tinker chemically with individual problems that pose no threat to society at large. Even if such gene-happy governments should emerge, evidence abounds that, in this country at least, many influential observers on both sides of the political fence stand ready to pounce at the first sign of overly eager gene manipulators.

The potential is also arriving to address directly through gene therapy not just some mental illnesses but disadvantageous aptitudes as well. Sandra Scarr is just one of many psychologists who foresee the possibility of raising I.Q. at the biochemical level as the genetic pathways to cognitive development become more thoroughly understood. This, she feels, is particularly auspicious in the case of various forms of mental retardation that have not responded to educational treatment. The potential applications for everyone else are no less exciting. We might envision the perfect Christmas gift of the future: fifteen more I.Q. points.

Another benefit to broad acceptance of the genes-behavior dynamic could be greater acceptance of human differences. As more and more people come to accept a biochemical basis for behavioral traits, the result could be an increased tolerance of the foibles of those around us. Although, as Ruth Hubbard and a number of opponents of behavioral genetics correctly point out, establishing a genetic basis for a condition is no guarantee of warm acceptance—as racial minorities and people with physical disabilities know well.

Still, it is not unreasonable to hope that those who sit in judgment of their fellow humans, which means all of us, will eventually make a distinction between a trait that stems from genes rather than one that is a choice. I have seen this work in my own case. I had always had a strong dislike of extremely effeminate men, especially the drag queens currently in fashion. But when I read Chandler Burr's *Atlantic Monthly* article "The Biology of Homosexuality" and learned of the hormonal mishaps in the womb that appear to cause femininity in men and masculinity in women, my hostility evaporated. I no longer saw outrageous queens as mincing embarrassments to all gays but rather as a biochemically betrayed group who make the best of the genetic hand they were dealt by being amusing (and looking fabulous).

Genetic awareness, while not ensuring toleration, at least will force the judgmental to find other pretexts for condemnation. In the case of homosexuality, even the most resolute homophobe should recognize that a genetic-biological cause is a sizable distance from morality-freighted environmental explanations, such as children yielding to infantile lusts that more strong-willed types were able to withstand. Perhaps in the coming genetic-awareness utopia, the gay-hater will acknowledge a nucleic basis, not just to gayness, but to his own *revulsion* at gayness. Such a perception, in turn, might lessen talk about the laws of God and of Nature.

ALWAYS LURKING NEARBY when one discusses applications of genetic knowledge is, of course, the much feared eugenic approach to inherited afflictions. While still a scare term, eugenics is already being practiced in America on a do-it-yourself basis by, among others, the Ashkenazi Jews, who, before marrying, often screen their partner's DNA for the life-destroying Tay-Sachs disease that plagues their group. If one of the couple has the dreaded gene, they frequently decide against marriage or against having children. But the ethics and morality of lethal diseases are easy. As scientific capability expands to include undesirable personality traits, the ethical dilemmas burgeon.

Within a few years, science will be able to predict in fetuses a high likelihood of such unwanted behavioral traits as depression, addiction, crippling timidity, violence. Although no one envisions forecasts of 100 percent certainty, merely a high probability, a recent poll dramatized the worrying repercussions of even this partial predictability. A group of young couples were asked if, knowing that their fetus has a fifty-fifty chance of becoming an obese adult, they would abort. Over three-quarters said they would. If such thinking is typical, Kate Smith, Gertrude Stein, or Luciano Pavarotti's chances for survival would have been, well, slim; they might have been sacrifices on the altar of genetic perfection.

This raises a hopeful point about the whole idea of at-home eugenics. Science is now drawing closer to a cure to the genetic anomaly that causes one form of obesity, obviating the abortion temptation. There is no reason for even the most finicky parents to abort a Rosanne or a John Goodman if we have the ability to make them thin later. In the coming years, similar correctives for genetic flaws will surely arrive for many behavior and trait problems. But until such remedies provide easy solutions to grim problems, the predictive potential of genetic screenings of fetuses presents enormous moral dilemmas and are laid out in detail in biogeneticist Philip Kitcher's book *The Lives to Come*.

But until the gene-adjusting cures are here, aborting fetuses marked for calamity remains a feasible option. Even for deadly diseases like Tay-Sachs, however, this would be eugenics pure and simple—but they are not the state-imposed eugenics so rightly feared by many. On the other hand, few of those for whom the word *eugenics* means Hitler and state-sanctioned genocide would condemn parents who decide against dedicating their futures to an imbecilic or schizophrenic offspring.

Even before coming close to the inhumanity of the Nazis or the superficiality of the fat-fearing parents, the moral dilemmas are overwhelming. And these will burgeon as science moves beyond the mere ability to predict the high likelihood of a trait to the ability to adjust, even eliminate, it. Would Oscar Wilde have been a great wit without his homosexuality, or Peter the Great the modernizer of Russia without his violent streak?

Brooding about this ability to eliminate flaws, genetic psychiatrists specializing in depression tell a dark joke about Handel's having written *The Messiah* in a two-week frenzy of manic elation. Had lithium been around at the time, they say, his symptoms might have been cured. As we approach the capability to edit genes in the womb, the result might be to strip the world of future Handels, Dostoyevskys, and van Goghs in a drive to prevent the birth of flawed humans. Another point in the debates to come would be that having a van Gogh on the planet was great for the rest of us but wasn't so great for van Gogh. Whose agony is it, anyway?

Conclusions

Books have and will be written about the moral and ethical conundrums that will ensue from the unraveling of human biochemical mysteries. As with any new discoveries, however, it should be kept in mind that knowledge, no matter how frightening, in itself, is neutral. Decisions about the *use* of knowledge immediately involves values, but society at large, not science alone, will have to grapple with those decisions. Still, with the dawning genetic and molecular biological knowledge, for all the perils and hard choices, few would deny its enormous potential for reducing human misery.

APART FROM SCREENING techniques that can now avert birth disasters and the eventual genetic corrections of potential problems, much good has already been accomplished by the simple recognition of genetic links to human behavior. This is most apparent in the area of mental illness, with attention at last being directed to biological causes of conditions long thought psychodynamic. As one genetic psychiatrist put it, using talk therapy on some forms of mental illness is like trying to talk a patient out of a kidney disorder. Freud himself saw the potential in biochemical knowledge when he wrote: "The shortcomings of our description would probably disappear if for the psychological terms we could substitute physiological or chemical ones. . . . Biology is truly a realm of limitless possibilities."

The new genetic perception has a potential for dispelling guilt on both the part of those with behavioral problems and on the part of parents who, in the environmental paradigm, have been wrongfully accused of causing it. One's heart goes out to the couple who stares at the floor as a therapist explains that their lack of love and support for their son has turned him into the depressed addict they now confront. Often, they are too cowed by the degree-holder before them and too ashamed of their inevitable parental failings to cite *less* supportive parents whose kids turned out okay. Recognition of a genetic component does not explain, much less exonerate, all behavioral dysfunctions, but it should lead to more realistic efforts to alter or remedy them. It should also deter the glib indictments of parents and others in the vicinity.

. . .

WHAT ARE THE BENEFITS of behavioral genetics perceptions for the majority of people who are not retarded, mentally ill, or a member of a despised genetic minority? What difference does it make to know that our DNA is telling our RNA to tell our ribosomes to make proteins that will influence our behavior? As long as all parties are doing their job and not mandating a problem-plagued life, of what use is this new understanding?

An important answer is that it can overhaul the self-view of every individual. And alter it in ways that could be useful and beneficial, not just provide a new framework for self-contemplation like astrology and other let's-talk-about-me quackeries. (You are a sensitive, feeling person for the following reasons . . .) It should change the way each of us thinks about our wants, our decisions, our emotions, our responses— change the way we view all the dynamic mechanisms that make us who we are. It might also have the humbling effect of making us less quick to assume a purely rational basis to our every thought, action, and emotion. It should make us skeptical of the facile rationales we assign to behaviors other parts of our brain have already ordained—for unconsidered or unreachable reasons. It should foster a hard-nosed skepticism about our choices and judgments and make us alert for atavistic, perhaps gene-propelled, impulses like racism and sexism that we don't want and that make no sense in today's world.

In contemplating human events that don't touch us directly, a genes-behavior awareness might render us more curious about underlying causes rather than cultural-historical circumstances. This would be particularly useful when we are rocked by stupefying examples of human violence—the Holocaust, Stalin's eradication of Ukrainians by the millions, the Khmer Rouge's slaughter of a million Cambodians. Invariably our reaction has been astonishment and horror at particular installments of humanity's periodic mass killings. If we reflect, we ponder the murderers' rise to power, their punishment, their present whereabouts. In this century, we also show interest in the technology that permits such large numbers of corpses in such a short time— Dresden and Hiroshima, for example.

With the genes-behavior link in mind, we might now also direct our attention toward the one element common to all of these sorry events: the aspects of human nature that permit such atrocities with a regularity we prefer to ignore. We should look beyond environmental exigencies and killing techniques and reserve at least some of our curiosity for the neuronal receptors and latent genetic impulses that enable these perpetual eruptions of homicidal frenzy.

For all our species' dazzling evolutionary success, we have carried with us from Neolithic times some ugly and counterproductive genetic baggage. As Robert Wright said in *The Moral Animal: Evolutionary Psychology and Everyday Life,* most evolutionary psychologists agree that natural selection does not aim for the overall good of the group and that the nature we humans have inherited is heavily freighted with ruthless self-interest. Some evolutionary thinkers (George Williams in *Adaptation and Natural Selection*) even go so far as to see natural selection as intrinsically evil, a built-in enemy we must continually resist in order to achieve a moral life. For this conflict to exist at all, an innate drive for good must exist within us as well, so it would follow that we must continually battle portions of our own natures to make a viable and just society.

This model merely reframes old paradigms about human duality—original sin, Plato's conflicting winged horses, man's moral ambiguity, and so on. But if the evolutionary psychologists are right about an inherited ruthless component to human nature, it adds a deeper understanding to the perception of the human predicament and, if the theory is confirmed by the evidence, would inject a large dose of empirical reality to concepts that had previously only been intuited and expressed in allegorical and sometimes lovely myths and belief systems.

It is not necessary to see ourselves as walking good-evil battlefields, Milton's "plains of heaven," but just as collections of countless genetic impulses—some good, some evil, some depressing, some joyous, some altruistic, some selfish, some prudish, some bawdy, some sad, some hilarious, some wise, some silly—all those unconnected neurons that we call a personality firing at different times and with different intensity in each of us, the human variation that makes for great novels and plays. The genetic cacophony can also produce searing inner conflicts and total breakdowns.

Still, it is unnecessary to see your DNA as an enemy, merely as a tricky friend, or more aptly, a battery of friends, some with considerable, but not total, powers of manipulation. Instead of brooding about the relative power of this or that nucleic nudge, however, we should be heartened by the simple fact that whatever power genes have to influence our behavior, that power is greatly diminished by an awareness of their incessant lobbying.

Because most of us are masters at dressing up genetic impulses with logical rationales, it is not always easy to recognize neucleotide interference in our thought processes. A spiteful remark makes you angry, for instance, or a friend's good news makes you envious and hostile. Such responses may spring from unadulterated thought or from genetic wiring; it is not easy sorting out which is which. It would be nice if science would develop a pocket brain monitor that buzzed each time our thinking was invaded by a reptilian instinct. But until such a device comes along, the best we can do is to develop a high level of suspicion about thoughts, desires, and actions that prove difficult to justify—even in terms of self-interest. A good place to start is with blatantly destructive and counterproductive emotions—like jealousy—over which most of us have no more control than we have over the pain felt when placing a hand over a flame. When attitudes and feelings trip us up or are hurtful to our interests, our pocket buzzers should go off. Knowing that a reaction stems from genes doesn't assure our controlling it, but we might hesitate to respect it and outfit it with the trappings of rationality.

As AWARENESS OF DNA over the details of our personalities spreads, it might throw light on the great debates that appear to be permanently embedded in modern society. The Minnesota Twin Study and other research projects have revealed genetic components to social attitudes, even political leanings. In the sentry's soliloquy in *Iolanthe*, W. S. Gilbert poked fun at the deep thoughts of a dull-witted man by having him sing,

> *Every boy and every girl*
> *That's born into the world alive*
> *Is either a little liberal*
> *Or else a little conservative.*

It now appears the sentry may have been right. In 1954, well after Gilbert but many years before the flowering of behavioral genetics, H. J.

Conclusions

Eysenck did a study with more than three hundred pairs of English twins that examined two broad personality factors: "radicalism" and "tough-mindedness." The former was based on left-right configurations of British politics; the latter was extrapolated from attitudes on corporal punishment and the death penalty. The heritability of both attitudes proved to be substantial.

Other studies have indicated that genes may influence, not just overall political leanings, but also rigidly held positions on such controversial issues as gun control, capital punishment, and gay rights. At first glance, such precise targeting of genetic influence may seem wildly improbable—or explainable by a broad genetic predisposition that leads, not so remarkably, to the specific stance. But before dismissing this notion of genetic underpinnings to specific attitudes, it should be remembered that the Minnesota study found even more specific behaviors and personality facets to have genetic origins. If the odd MZA identicalities have no other scientific utility, they can at least serve to raise the *possibility* of pinpointed genetic foundations in attitudes and opinions. If precise choice of hobbies, pet names, and wardrobe items can spring from a bit of DNA, as the reared-apart twin studies strongly suggest, the notion appears less far-fetched that genes might play a role in more fundamental aspects of an individual's persona, such as intense positions on contentious issues.

The most easily digestible findings in the area of social attitudes have been with broad, personality-defining tendencies that have societal ramifications. In 1981 Sandra Scarr found a degree of heritability for authoritarianism, an affinity for strict discipline. If such a trait can be transmitted genetically, it might explain why some individuals (or nations) thrive under authoritarian political and religious systems and others rebel against them.

A 1975 Australian study of 3,810 pairs of twins discovered a genetic component to a broad range of attitudes, everything from a liking for modern art to respect for divine law. Perhaps the most significant findings had to do with racial attitudes. Of three questions relating to this subject—belief in white superiority, acceptance of mixed marriages, and feelings about nonwhite immigration—all had a significant degree of heritability. These findings raise the intriguing possibility that the racists who reveled in *The Bell Curve*'s genetic allegations about I.Q.

winners and losers may themselves be in the genetic grip of some unattractive evolutionary residue.

While indulging in such freewheeling speculations about possible genetic underpinnings to social attitudes, it should be remembered that hard evidence of gene involvement says zero about a trait's desirability or moral value. It also says zero about its immutability or its power over individuals. On the plus side, the mere awareness of possible genetic foundations for strong opinions holds out hope for mitigating venerable antagonisms.

Should, for instance, a biochemical basis be clearly established in our attitudes toward such stubborn problems as racial hostilities or capital punishment, it might explain why the endless appeals to reason and fairness, from both sides, have historically had so little effect. Strong environmental influences—whether *New York Times* editorials, Pat Robertson sermons, or Aryan Nation propaganda—can, I suspect, reinforce the positions of those whose genes make them receptive to the ideology being pitched. I strongly doubt if such appeals can persuade people whose genes point them in a different ideological direction. Fortunately, many people are neutral on issues—perhaps in genetic balance—and it is they, the ever-popular undecided, that the political strategists go after.

Even if genes could be proven to tilt us one way or another on an issue, that is not to say we don't have other genes working to tilt us back again. One gene might underwrite a strong belief in harsh punishment of crimes yet from somewhere further down a chromosome another gene also prompts a revulsion at killing. On the death penalty issue, this would make for a genetic teeter-totter that a well-written editorial or a dinner-table argument could tilt one way or the other.

One of those most prolonged and intractable debates of contemporary society concerns abortion. Many today are gripped by the tragedy of neglected children and other grim consequences of unwanted pregnancies. From the evolutionary point of view, by contrast, it is not difficult to see an adaptive advantage—in an ancient planet with few humans and many predators—to a revulsion at killing fetuses becoming an element of human hard wiring. It could be one entry in our catalog of inherited behaviors, probably a variant of the instinct for protecting offspring seen in most species, and it could be stronger in some individuals than others.

Although admittedly a freewheeling conjecture, if a horror of abortion, in some people, has lingering genetic roots, it might throw light on the imperviousness to argument of most abortion opponents. It could suggest a reason why for so many decades of heated debate the opposing sides have not heard each other. The pro-choicers' reasons—maternal rights, children's welfare, benefits to society—could be colliding with a gene in a portion of the population that says simply, "Killing fetuses is wrong." It is, to me, conceivable that this law, in those who see it as bedrock truth, may have been reinforced by religious indoctrination, ethnic tradition, or moral persuasion but stemmed originally from a gene.

While contemplating the possibility of genes influencing attitudes and opinions, it is useful to remember the genes-environment mix-ups discussed in chapter 9. Do book lovers come by their passion through upbringings in book-filled homes or did they inherit a book-loving gene from their parents? Abortion opponents may have come by their aversion through religious upbringings, or their antiabortion stands as well as those of their parents may result from a family gene, one that possibly combined with other "morality" genes to make them *and* their parents religious.

If some distant scientific discovery could prove genetic foundations to such unbending positions, where would it leave us? Nowhere, as far as morality is concerned, since genetic impulses span the good-evil gamut. If, however, an individual's racial hatreds or another's fervor for animal rights could be shown to be genetic, this might, just might, make them less dogmatic, more receptive to argument. A genetic element in no way means that we can't reject such feelings, only that should our intellects decide against them, we would have to work harder than others to overcome them.

Evidence of genetic links to attitudes does not sanction a passive resignation to one's makeup. It certainly does not imply absolution. To shrug one's shoulders in this way assumes an impossibility to countermanding genetic impulses that is a myth I hope this book has put to rest. We still have the ability to sift through the endless blips and flashes emanating from our DNA, go with those we approve, and reject those we dislike. In fact, it is the ones that seem to come from nowhere, that clash with our self-view, that should be most guarded against, viewed with the highest skepticism. Few of us haven't struggled against an unwanted attitude, convinced ourselves that reason decrees we must abandon it, only to have it snap back when we stop thinking about it. A distaste for physically deformed people might be an example, or a racist attitude that our reason says is uncivilized and unfair. I also suspect the attitudes we cling to most tenaciously, the causes and beliefs that inspire the highest passions (or the ones we most vehemently deny have any connection with genes), are prime suspects for emanating from the genome rather than the intellect—if only because of their power over us.

Militants on either side of intractable social issues would surely welcome a silver-bullet pill, a gene-altering chemical, that would bring about conversions in their opponents. But it is unnecessary to conjecture molecular magic wands for resolving long-running arguments to foresee benefits to establishing a genetic component to these battles, should science arrive at such a conclusion. The benefit most easily visualized would be a softening of stances on both sides of public debates. With abortion, for example, those believing in reproductive choice might see their adversaries not as religious fanatics or brainwashed extremists but as people impelled by strong signals from their hard wiring. The antiabortionists, in turn, might see their opponents not as baby murderers but as individuals lacking this particular genetic dictate. Either view, in my opinion, points, if not to acceptance of opposing stances, at least to greater patience, perhaps tolerance.

There is an even more ambitious hope. If the pro-lifers could be persuaded that their revulsion has a genetic foundation, this might tend to disconnect their position from divine law, church dogma, or other rationales beyond the reach of argument (or, in the eyes of a few, beyond the reach of law). Genetic impulses, at times, can also be impervious to argument. You can't argue yourself out of hunger, for example. But if antiabortionists, or any cause crusaders (whether "right" or "wrong"), could see their stands as grounded in their biochemical makeup, rather than as moral convictions swathed in a religious majesty, they might be more amenable to persuasion and compromise. Nucleotide nudges can be overruled; God's law cannot.

Such a revolution in thinking about our opinions would surely require hard proof of genetic foundations of attitudes; to date, such a connection has been little more than suggested by the evidence. Behavioral geneticists are currently more concerned with acute personality problems like addiction, depression, and mental illness. Still, even the possibility of specific genetic links between genes and attitudes, to my mind, makes this line of genetic research particularly auspicious.

STARTLING RESEARCH ON brain function conducted by Michael Gazzaniga, the director of the University of California's Center for Neurobiology, promises to deepen understanding of gene-based behavioral impulses. It also has implications for group actions, even the epic events of history. In elaborate experiments with patients with severed brain hemispheres, Gazzaniga and colleagues have learned much about left-right functioning and about the numerous brain modules, perhaps thousands—configurations of neurons that seem to be operating independently like little minds. (If you think this is hard to swallow, try the belief of N.Y.U.'s Rodolfo Llinas, one of the world's top brain experts, who believes each of the brain's billions of neurons has a mind of its own.) Because each eye is hooked up to a different hemisphere, and because the damage prevents the two sides from communicating as usual, the neuroscientists are able to "talk" to just one hemisphere in these individuals by covering an eye and holding up written signs. In this way, a lot has been discovered about the origins and mechanics of mental processes.

Most interestingly, the scientists have located a center in the left hemisphere that processes and makes sense out of the impulses and messages constantly flowing from all other areas of both hemispheres— including the gene-based behavioral signals that are the subject of this book. Gazzaniga calls this region of the brain "the interpreter," and it turns out to be quite a remarkable, resourceful, and not altogether honest aspect of our cognitive apparatus. In the experiments, testers gave instructions to the right side of a subject's severed brain, then asked the left or "interpreter" side why the subject performed the requested action. Ignorant of the real reason, the researchers' input, the subject's interpreter had no problem ad-libbing a rationale.

An example: The researchers held up a sign to one side of the subject's brain saying WALK. When addressing the brain's other side, the subject rose from his chair midtest and began to walk away. Asked what he was doing, he hesitated, then replied, "I'm going to get a Coke."

Another subject was shown a sign that said LAUGH. Later, when using her brain's other half, she began to chuckle. Asked what amused her, she answered that she found it funny that people made livings asking such questions.

Gazzaniga describes the interpreter as a kind of all-purpose public relations person, a front-office spokesman for all the independent brain modules. Its job is to explain and make sense of actions decided on by other brain segments—for reasons of which it is often ignorant. At first glance this seems a role similar to that of the White House press officer, who must put the most plausible public face on decisions he or she had no part in making and the real reasons for which, if he knows them at all, must not be revealed. But the analogy misses a major aspect of the interpreter's role in our brains; it must deceive not only the outside world but must also deceive ourselves. Unlike the White House press officer, our brain's interpreter believes what it says, no matter how far-fetched. So in terms of gene-based behavior, a bit of nucleic acid prompts an action; if the signal encounters no counter-signals as it speeds along the nerve network, it enlists the brain's interpreter to provide as reasonable a rationale as possible for the action that it can muster, and the action occurs—armed with an explanation in case one is needed by others or by the individual taking the action.

The implications are enormous for this human mechanism for invented explanations and off-the-cuff pretexts. When it is applied to the conclusion of behavioral geneticists—that we are all subjected to a continual barrage of biochemically delivered moods, nudges, shoves, impulses, craving, and aversions—it suggests that however bizarre, hurtful, or inappropriate one of our actions or mental responses may be, the interpreter module in our brain stands ready to provide it with a reasonable face. We are perfectly capable of saying, "I'm going to get a Coke," when we actually have no idea of what we are doing or why. Of course, sometimes we actually want a Coke, but other times we concoct such a reason because we need a *reason*, not a Coke.

As for the chances of seeing through the interpreter's smooth baloney, the higher one's intelligence, the more cunning one's interpreter probably is at internal hoodwinking. Another mechanism may further strengthen our ability to delude ourselves. The more distasteful one of our own impulses is to us—that is, the more in conflict it is with our nobler impulses—the more adept the interpreter may be at inventing rationales and pretexts. And to raise the potential for self-delusion yet another notch, the more moral and decent we perceive ourselves to be, the more we may put our interpreters to work grinding out pretexts when those immoral and indecent genetic impulses, our caveman legacy, clamor for expression.

NOWHERE DOES THE CONCEPT of "pretext" take on greater resonance than in humankind's grim history of serial warfare. While our species' built-in human predilection for homicide was a major thesis of the influential writings of Lorenz and Ardrey, it has been dismissed by many scientists and members of the public who reject such an unappetizing view. The denial is remarkable in the face of the historical record and the number of wars raging on the planet at any given moment—to say nothing of the countless "private" murders and the billions being raked in by the pornography of violence: films, television, computer games, mystery thrillers, toy guns, and the countless other murder substitutes our society offers. Self-delusion on this ugly element of our species' basic equipment seems so widespread, it suggests millions of interpreters working full time to conceal the nasty truth about our homicidal leanings.

Genetically oriented psychologists are now zeroing in on another built-in behavioral mechanism that may well hook up with our innate aggressiveness to make humans as warlike as we are. Since the 1950s, researchers have been examining the speed with which humans, even very young humans, can form themselves into "us" and "them" groups. The investigators also have found how quickly these identities can spawn hostilities between the in-group and out-group. It might be an arbitrary dividing of third graders into Blue Jays and Cardinals or summer campers into Rattlers and Eagles. No matter how arbitrary the groupings, animosities quickly emerge that can escalate into hostility, even violence. The most casual glance at other cultures suggests a specieswide aspect to the phenomenon. It might be anything from Rwandan tribes to Russian ethnicities to Liverpool soccer fans. The speed with which this mechanism operates can be seen most horrifyingly in the Serbs and Croatians, who, within a few months, discovered they loathed and wished dead the group they had lived with peaceably for years.

If there is a genetic mechanism that makes us prone to identify with one group and hostile to another, it appears to be closely related, may in fact activate, a genetic mechanism for aggression. The same neurological parlay that converts normal humans into lynch parties and murderous mobs may be more benignly at work in nursery schools and fraternity houses. Whatever the premise for group identity, it doesn't take long for the out-group to go from being "different" to being inferior, unworthy, evil, contemptible—until they are seen as needing harassment, punishment, retribution, perhaps annihilation.

Among the few writers who have addressed this human gusto for identifying, then killing enemies, one of the most succinct was Francis Ford Coppola in his screenplay for *Patton*. After a particularly grizzly battle, George C. Scott, as Patton, surveys the killing field strewn with smoking tanks and bloody corpses. As the camera pans across the horror spread before him, audiences await the obligatory war-is-hell line that absolves Patton for his murderousness (and absolves viewers for paying to watch it). Instead they are jolted to hear Patton say, "I love all this. God help me, but I love it." While George Patton could be dismissed as a blood-thirsty anomaly, the same unsparing self-awareness was evinced by as disparate a figure as Marianne Moore when she wrote,

> . . . I must
> fight til I have conquered in myself what
> causes war . . .

So far, few have demonstrated this knack for harsh self-analysis. Instead of looking inside ourselves to explain human activities we abhor—crime, war, violence, oppression, worker exploitation—we invariably have at our disposal a catalog of environmental explanations—Hollywood, Wall Street, Madison Avenue, Gangsta Rap, the N.R.A., and, the old standbys, derelict parents. If we run out of present-day causes, we have the past to draw on. A dig into history is sure to unearth the root of trouble and pinpoint a historical reason one group or another has strayed from our species' alleged peaceful, nonviolent nature. (Good grief, are they doing it again?)

Recently, this reflex was most poignantly demonstrated with the Bosnian tragedy. In trying to determine why these former Yugoslavians found it necessary to uproot, kill, and torture one another, the *New York Times* doughtily dug back to the fourteenth century—to the battle of Kosovo in 1389. This event, it seems, festered for six hundred years, only to produce mass slaughter in the 1990s. Although the journalists are simply doing what tradition holds to be a thorough analysis, they rely on the assumption that explanations for such irrational behavior must lie *somewhere* in the environment, even a centuries-old environment.

I hear no one suggesting that the prior events may not be causes at all but may merely be earlier symptoms of the same built-in malaise— or, more likely, earlier symptoms of shared human behavior that erupts with disheartening regularity throughout history (when environmental conditions are right, to be sure). In addition to the possibility that the past-scanning analysis may miss the mark—be nowhere near it, in fact—the earnest efforts of the more serious journalists to locate historical reasons has the inadvertent effect of dignifying the present carnage, elevating it to the logical result of earlier events. Historical precedents are forced into service to make the insane appear sane, or, in my view, to make the genetic appear environmental. Once again, interpreters working big time.

One of the best analyses of the Bosnian horror was Peter Maass's 1996 book *Love Thy Neighbor,* in which the author refuses to credit the historical explanations for the killing and speaks instead of manipulations by evil leaders of "the wild beast" within us all. This, I feel, brings Maass closer to the truth than most war-watchers; he feels that historical antecedent is inadequate explanation of the genocidal rampage and looks instead for something in the makeup of all humans.

While I cheer this shift of focus, I also feel that in 1998 we can do better than such pallid literary metaphors as the "beast within us." In fact, we don't need metaphors at all. We are close to understanding the precise biochemical mechanisms that can, with apparent ease, lead us to declare a group the enemy and attack it. That aspect of human personality that engenders atrocities like Bosnia is not merely "like" something in raw, brute nature. It is an integral, if perhaps often dormant, part of *our* nature and one that behavioral geneticists are understanding better every day.

Two present-day developments hold out hope that humans may at last be able to see beyond environmental circumstances that are

271

constantly served up by history-minded journalists to explain ubiquitous warfare. One, of course, is the dawning genetic self-awareness that is the subject of this book. The other is the nonsensical nature of the current crop of conflicts. At no time in recent history has the planet seen more wars fought for such strained and implausible rationales. Perhaps at the moment there is a dearth of incontestable, morality-grounded reasons for fighting such as those that ennobled World War II. Then, too, perhaps the universal interpreter has grown careless or just run out of credible reasons for humans to kill other humans. But puzzling, baseless wars are proliferating as never before, and they appear to be as lethal as the high-minded ones.

In *Civil Wars*, a 1994 book of essays about the rash of senseless violence that has plagued the world since the end of the Cold War, Hans Magnus Enzenberger despairs that wars are no longer waged for righteous causes; instead, "violence has freed itself from ideology." What Enzenberger seems to see as an unfortunate development—where are the nifty wars of yesteryear?—I see as a rare opportunity for humans to perceive their aggressions for what they invariably are: ancient genetic impulses that aided survival at one time but now work against it. The rash of ill-founded wars may help us look beyond the pretexts of the moment to the underlying human mechanisms. The senseless wars may help us better understand the genetic underpinnings of the "sensible" ones. When about to march off for God or country, we would do well to consider the New Guinea highlanders who periodically attack their neighbors because they feel like it. We should also consider chimpanzees, our closest relatives, who regularly stage murderous raiding parties on neighboring tribes for no discernible reasons of hunger, sex, or territory.

The hope is that instead of swallowing without question the moment's rationales for killing, we might one day come to see ourselves, when in the warlike mode, in the grip of a genetic restlessness, perhaps a DNA surge of group aggression triggered by a circumstance or by a demagogue, an underlying motivation that renders our war no more rational than those of the New Guinea tribesmen. In this regard, the only difference between our two cultures may be that, unlike ourselves, the New Guineans feel no need for hypocritical, self-justifying rationales to kill others. They know they slaughter to relieve boredom and because they enjoy it. We love all this, God help us.

Conclusions

. . .

WHETHER OR NOT the young science of behavioral genetics finds the root cause of behaviors we deplore, like war and racism, this field of inquiry, along with its sister science of evolutionary psychology, is arriving at a broad new understanding of our species. In addition to seeing ourselves as products of culture, education, and upbringings, we can also see ourselves as blinking switchboards of gene-fired impulses, some older than the species itself, some weak, others powerful, some ever-present, others sporadic—but all waiting their moment to take charge of the entire vessel, to move us to an action that evolution at one period in our four-million-year history decided increased chances of survival.

After only two decades of concerted research into this aspect of our makeup, we can now address human dysfunctions, contradictions, and self-destructiveness armed with a grasp of an important new component, perhaps the most important of all: the powerful effect on behavior of the human genome, the twenty-three pairs of chromosomes that produced our eyes, feet, and kidneys, and play a role in every aspect of our behavior.

Appendix

The following letter appeared in the July 1972 issue of American Psychologist.

BEHAVIOR AND HEREDITY

The posthumous Thorndike Award article by Burt (1972) draws psychological attention again to the great influence played by heredity in important human behaviors. Recently, to emphasize such influence has required considerable courage, for it has brought psychologists and other scientists under extreme personal and professional abuse at Harvard, Berkeley, Stanford, Connecticut, Illinois, and elsewhere. Yet such influences are well documented. To assert their importance and validity, and to call for free and unencumbered research, the 50 scientists listed below have signed the following document, and submit it to the APA:

BACKGROUND: The history of civilization shows many periods when scientific research or teaching was censured, punished, or suppressed for nonscientific reasons, usually for seeming to contradict some religious or political belief. Well-known scientist victims include: Galileo, in orthodox Italy; Darwin, in Victorian England; Einstein, in Hitler's Germany; and Mendelian biologists, in Stalin's Russia.

Today, a similar suppression, censure, punishment, and defamation are being applied against scientists who emphasize the role of heredity in human behavior. Published positions are often misquoted and misrepresented; emotional appeals replace scientific reasoning; arguments are directed against the man rather than against the evidence (e.g., a scientist is called "fascist," and his arguments are ignored).

A large number of attacks come from nonscientists, or even antiscientists, among the political militants on campus. Other attackers include academics committed to environmentalism in their explanation of almost all human differences. And a large number of scientists, who have studied the evidence and are persuaded of the great role played by heredity in human behavior, are silent, neither expressing their beliefs clearly in public, nor rallying strongly to the defense of their more outspoken colleagues.

The results are seen in the present academy: it is virtually heresy to express a hereditarian view, or to recommend further study of the biological bases of behavior. A kind of orthodox environmentalism dominates the liberal academy, and strongly inhibits teachers, researchers, and scholars from turning to biological explanations or efforts.

RESOLUTION: Now, therefore, we the undersigned scientists from a variety of fields, declare the following beliefs and principles:

1. We have investigated much evidence concerning the possible role of inheritance in human abilities and behaviors, and we believe such hereditary influences are very strong.
2. We wish strongly to encourage research into the biological hereditary bases of behavior, as a major complement to the environmental efforts at explanation.
3. We strongly defend the right, and emphasize the scholarly duty, of the teacher to discuss hereditary influences on behavior, in appropriate settings and with responsible scholarship.
4. We deplore the evasion of heredity reasoning in current textbooks, and the failure to give responsible weight to heredity in disciplines such as sociology, social psychology, social anthropology, educational psychology, psychological measurement, and many others.
5. We call upon liberal academics—upon faculty senates, upon professional and learned societies, upon the American Association of University Professors, upon the American Civil Liberties Union, upon the University Centers for Rational Alternatives, upon presidents and boards of trustees, upon departments of science, and upon the editors of scholarly journals—to insist upon the openness of social science to the well-grounded claims of biobehavioral reasoning, and to protect vigilantly any qualified faculty members who responsibly teach, research, or publish concerning such reasoning.

We so urge because as scientists we believe that human problems may best be remedied by increased human knowledge, and that such increases in knowledge lead much more probably to the enhancement of human happiness than to the opposite.

Signed:

JACK A. ADAMS
Professor of Psychology
University of Illinois

DOROTHY C. ADKINS
Professor/Researcher in Education
University of Illinois

ANDREW R. BAGGALEY
Professor of Psychology
University of Pennsylvania

IRWIN A. BERG
Professor of Psychology and Dean of Arts Sciences
Louisiana State University

EDGAR F. BORGATTA
Professor of Sociology
Queens College, New York

ROBERT CANCRO, MD
Professor of Psychiatry
University of Connecticut

Appendix

RAYMOND B. CATTELL
Distinguished Research Professor
   of Psychology
University of Illinois

FRANCIS H. C. CRICK
Nobel Laureate
Medical Research Council
Laboratory of Molecular Biology
Cambridge University

C. D. DARLINGTON, FRS
Sherardian Professor of Botany
Oxford University

ROBERT H. DAVID
Professor of Psychology and Assistant Provost
Michigan State University

M. RAY DENNY
Professor of Psychology
Michigan State University

OTIS DUDLEY DUNCAN
Professor of Sociology
University of Michigan

BRUCE K. ECKLAND
Professor of Sociology
University of North Carolina

CHARLES W. ERIKSEN
Professor of Psychology
University of Illinois

HANS J. EYSENCK
Professor of Psychology
Institute of Psychiatry
University of London

ERIC F. GARDNER
Slocum Professor & Chairman
Education and Psychology
Syracuse University

BENSON E. GINSBURG
Professor & Head, Biobehavioral Sciences
University of Connecticut

GARRETT HARDIN
Professor of Human Ecology
University of California, Santa Barbara

HARRY S. HARLOW
Professor of Psychology
University of Wisconsin

RICHARD HERRNSTEIN
Professor & Chairman of Psychology
Harvard University

LLOYD G. HUMPHREYS[1]
Professor of Psychology
University of Illinois

DWIGHT J. INGLE
Professor and Chairman of Physiology
University of Chicago

ARTHUR R. JENSEN
Professor of Educational Psychology
University of California, Berkeley

RONALD C. JOHNSON
Professor & Chairman of Psychology
University of Hawaii

HENRY F. KAISER
Professor of Education
University of California, Berkeley

E. LOWELL KELLY
Professor of Psychology & Director, Institute of
   Human Adjustment
University of Michigan

JOHN C. KENDREW
Nobel Laureate
MRC Laboratory of Molecular Biology
Cambridge, England

FRED N. KERLINGER[1]
Professor of Educational Psychology
New York University

WILLIAM S. LAUGHLIN
Professor of Anthropology & Biobehavioral
   Sciences
University of Connecticut

DONALD B. LINDSLEY
Professor of Psychology
University of California, Los Angeles

QUINN MCNEMAR
Emeritus Professor of Psychology, Education,
   and Statistics
Stanford University

PAUL E. MEEHL
Regents Professor of Psychology and Adjunct
   Professor of Law
University of Minnesota

[1] In item 1, preferred "substantial" or "important" to the wording "very strong."

Appendix

JACQUES MONOD
Nobel Laureate
Professor, Institute Pasteur
College de France

JOHN H. NORTHRUP
Nobel Laureate
Professor Emeritus of Biochemistry
University of California and Rockefeller
    University

LAWRENCE I. O'KELLY
Professor and Chairman of Psychology
Michigan State University

ELLIS BATTEN PAGE
Professor of Educational Psychology
University of Connecticut

B. A. RASMUSEN
Professor of Animal Genetics
University of Illinois

ANNE ROE
Professor Emerita, Harvard University &
    Lecturer in Psychology
University of Arizona

DAVID ROSENTHAL
Research Psychologist and Chief of Laboratories
National Institute of Mental Health

DAVID G. RYANS
Professor & Director
Educational R & D Center
University of Hawaii

ELIOT SLATER, MD
Professor of Psychiatry and Editor
British Journal of Psychiatry
University of London

H. FAIRFIELD SMITH
Professor of Statistics
University of Connecticut

S. S. STEVENS
Professor of Psychophysics
Harvard University

WILLIAM R. THOMPSON
Professor of Psychology
Queens University, Canada

ROBERT L. THORNDIKE
Professor of Psychology and Education
Teachers College
Columbia University

FREDERICK C. THORNE, MD
Editor, *Journal of Clinical Psychology*
Brandon, Vermont

PHILIP E. VERNON
Professor of Educational Psychology
University of Calgary, Alberta

DAVID WECHSLER
Professor of Psychology
N.Y.U. College of Medicine

MORTON W. WEIR
Professor of Psychology and
    Vice-Chancellor
University of Illinois

DAVID ZEAMAN
Professor of Psychology and NIMH Career
    Research Fellow
University of Connecticut

REFERENCE

BURT, c. Inheritance of general intelligence.
    American Psychologist, 1972, 27, 175–190.

ELLIS B. PAGE
University of Connecticut

Notes

PREFACE

ix "The reviews . . . were mostly negative": Interview with Michael
   Bessie, June 24, 1997.

x "when I read an article": *Smithsonian*, October 1980.

ONE: THE CHEMISTRY OF SELF

5 "one child in thirty": *The Language of Genes* by Steve Jones, Anchor
  Books, 1994.

8 "the front page of": *New York Times*, January 1, 1996.
  "a third study in Finland": *New York Times*, November 1, 1996.
  "a typical example of the misunderstanding . . .": *Time*, January 12, 1994.

10 "a gene in female rodents was found": *New York Times*, November 2,
   1993.

11 "tells a story of identical brothers": *Nature's Thumbprint* by Peter B.
   Neubauer and Alexander Neubauer, Addison-Wesley, 1990.

12 "The liberal movements that flourished": *In Search of Human Nature*
   by Carl Degler, Oxford University Press, 1990.

13 "Darwin had barely enunciated his theory": Ibid.
   "genes-behavior theories make sporadic appearances": Interview with
   Leon Kamin, September 21, 1993.

14 "critics pin this belief": *Not in Our Genes* by Richard Lewontin, Leon
   Kamin, and Steven Rose, Pantheon, 1985.

15 "when humans use contraception": *The Selfish Gene* by Richard
   Dawkins, Oxford University Press, 1976.

16 "this is like the man who lost his wallet": Interview with Robert
   Plomin, November 17, 1993.

17 "their many *differences* are eloquent testimony": *Why Children
   Are So Different* by Judy Dunn and Robert Plomin, Basic Books,
   1992.

17 "while the environment *can* have a major effect": "Personality Similarity in Twins Reared Apart and Together" by Thomas Bouchard et al., *Journal of Personality and Social Psychology*, vol. 54, no. 6, 1988.

18 "required a four-letter word": *New York Times Magazine* crossword puzzle, October 30, 1994.
"he launched into a slash-and-burn": Interview with Leon Kamin, September 21, 1993.

TWO: BIRTH OF A STUDY

22 "the only American separated-twin experiment": *A Study of Heredity and Environment* by H. H. Newman, F. N. Freeman, and K. J. Holzinger, University of Chicago Press, 1937.

23 "culminated in a 1981 book": *Identical Twins Reared Apart* by Susan Farber, Basic Books, 1981.
The information about the Cyril Burt case comes from two books: *The Burt Affair* by Robert B. Joynson, Routledge, 1989; and *Science, Ideology and the Media* by Ronald Fletcher, Transaction Publishers, 1991.

24 "Bouchard knew well": Interviews with Thomas Bouchard, November 6–12, 1993.

25 Some of the information about the Jim twins comes from interviews at the University of Minnesota; from *Smithsonian*, October 1980, and from *Twins* by Peter Watson, Hutchinson and Co., 1981.

33 "one of the day's most widely used psychology textbooks": *Introduction to Personality* by W. Mischel, Holt, Rinehart and Winston, 1981.

THREE: DESPERATELY SEEKING TWINS

35 Bouchard adopted a policy: "The Minnesota Study of Twins Reared Apart: Project Description" *Twin Research 3: Intelligence, Personality and Development*, Alan R. Liss, Inc. 1981.

36 " 'Only Tom Bouchard' ": Interview with Auke Tellegen, November 7, 1993.
" 'She's sitting in my living room,' " *Los Angeles Times*, April 18, 1992.

38 "Bouchard estimates": Interview with Thomas Bouchard, November 11, 1993.
"Bouchard's findings were later replicated": "I.Q. Similarity in Twins Reared Apart" by Thomas Bouchard, *Intelligence: Heredity and Environment*, Cambridge University Press, 1994.

40 *"every trait they measured* showed at least" "Source of Human Psychological Differences" by Thomas Bouchard et al., *Science,* October 1990.

41 " 'It's not like working with' ": Interview with Thomas Bouchard, November 15, 1993.

42 "he would get periodic updates": Interview with Arlen Price, October 6, 1993.

43 "developed the concept of *nonshared environment*": *Why Children Are So Different* by Judy Dunn and Robert Plomin, Basic Books, 1992.

44 "organisms go a long way toward creating their own environment": *The Extended Phenotype* by Richard Dawkins, W. H. Freeman, 1982.

FOUR: COSMIC SECRETS OF TWINS

47 "The head of the Pioneer Foundation": Telephone interview with Harry Weyher, June 6, 1993.

48 "Bouchard contributed a chapter": *Individuality and Determinism,* edited by Sidney W. Fox, Plenum Publishing, 1984.
" 'there is no compelling evidence' ": *The Science and Politics of I.Q.,* Lawrence Erlbaum Associates, 1974.

50 "published a paper on homosexuality": "Homosexuality in Twins Reared Apart" by Thomas Bouchard et al., *British Journal of Psychiatry,* vol. 148, 1986.

51 "far more comprehensive study of homosexuality": "A Genetic Study of Male Sexual Orientation" by J. M. Bailey and R. C. Pillard, *Archives of General Psychiatry,* vol. 48, December 1991.

FIVE: MINNESOTA'S TRIUMPHS

52 "the Minnesota group had completed assembling its data": *New York Times,* December 2, 1986.
"Good news for the environmentalists": "Personality Similarity in Twins Reared Apart and Together" by Thomas Bouchard et al., *Journal of Personality and Social Psychology,* vol. 54, no. 6, 1998.

55 "The paper, published in 1990": "Sources of Human Psychological Differences: The Minnesota Study of Twins Reared Apart" by Thomas Bouchard et al., *Science,* October 12, 1990.

58 "the magazine publishing only two critical letters": *Science,* April 2, 1991.

59 "In a later paper": "Twins: Nature's Twice-Told Tale" by Thomas Bouchard, 1983 *Yearbook of Science and the Future*, Encyclopedia Britannica.
"more fully developed by Harvard's Richard J. Herrnstein": *I.Q. in the Meritocracy* by Richard Herrnstein, Atlantic–Little Brown, 1973.

SIX: TWO DOGS NAMED TOY

62 "were expressed in figures of heritability": Most clearly defined on page 232 of *Behavioral Genetics: A Primer* by R. Plomin, J. C. Defries, and G. E. McClearn, W. H. Freeman and Company, 1980.

63 "One of the most interesting pairs of twins": Interviews with Minnesota staff and from *Twins* by Peter Watson, Hutchinson and Co., 1981.

67 "The twins were born in Trinidad": Most information about Jack and Oskar comes from telephone interviews with Jack Yufe, October 22 and November 2, 1993, and with Mona Yufe, October 13, 1993.

71 "Jack and Oskar remained stiff with each other": Interviews with the Minnesota staff, November 1979.

73 "46 percent of people polled": *New York Times*, April 22, 1994.

75 "twin differences might turn out to be more interesting": Interview with Thomas Bouchard, November 12, 1993.

SEVEN: MORE WEIRDNESS

78 "(twins who were both bachelor firemen)": *Smithsonian*, November 1979.

79 "The paper was based on fifteen pairs": "Preliminary Findings of Psychiatric Disturbances and Traits" by E. Ekert, L. L. Heston, and T. L. Bouchard, *Intelligence, Personality and Development*, Alan R. Liss, Inc., 1981.

81 "In his book": *Nature's Mind* by Michael Gazzaniga, Basic Books, 1992.

83 "David Lykken worked out a theory": "The Concept of Emergenisis," presidential address delivered at the annual meeting of the Society for Psychophysiological Research in Washington, October 31, 1981.

EIGHT: OTHER BEHAVIORAL GENETICS STUDIES

86 "the two scientists, along with Robert Plomin": *Behavioral Genetics: A Primer* by R. Plomin, J. C. Defries, and G. E. McClearn, W. H. Freeman and Company, 1980.

88 "Among the intriguing findings": "Temperament, Emotion and Cognition at Fourteen Months" by Robert N. Ende et al., *Child Development*, vol. 63, 1992.
"In a preliminary 1992 paper": Ibid.

89 "but was colored by strong altruistic ambitions": Interview with Jerome Kagan, September 21, 1993.

90 "For Kagan this faith began to weaken": "Cross-Cultural Perspectives on Early Development" by Jerome Kagan and Robert E. Klein, *American Psychologist*, November 1973.

91 "His epiphany came fifteen years later": "Biological Basis of Childhood Shyness" by J. Kagan, J. Reznick, and N. Snidman, *Science*, April 8, 1988.

92 "His major conclusion was stated in a 1988 paper": Ibid.

93 "In his 1994 book": *The Nature of the Child* by Jerome Kagan, Basic Books, 1994.

95 "Kendler became one of a growing number of American psychiatrists": Interview with Kenneth Kendler, March 12, 1993.
"In Ming Tsuang's 1997 book": *Schizophrenia: The Facts* by Ming Tsuang, Oxford, 1997.
"a study of bulimia": "The Genetic Epidemiology of Bulimia Nervosa" by K. Kendler et al., *American Journal of Psychiatry*, December 1991.
"and a study of sleep disorders": "Evidence for Genetic Influences on Sleep Disturbances and Sleep Pattern in Twins" by K. Kendler, L. J. Eaves, N. G. Martin, *Sleep*, vols. 13 and 14, 1990.
"The studies were based on female reared-together twins": "Generalized Anxiety Disorder in Women" by K. Kendler, *Archives of General Psychiatry*, April 1992.

96 "Similar results were reported": Ibid.
"The Virginia group also published a paper ": "The Genetic Epidemiology of Phobias in Women" by K. Kendler et al., *Archives of General Psychiatry*, April 1992.

97 "Kendler's Virginia group came out with a paper": "A Population-Based Twin Study of Alcoholism in Women" by K. Kendler et al., *Journal of the American Medical Association*, October 14, 1992.
"The far larger number of reared-*together* twins needed": *Behavioral Genetics: A Primer* by R. Plomin, J. C. Defries, and G. E. McClearn, W. H. Freeman and Company, 1980.

99   "is the concept of children molding their own environments":
"Distinctive Environments Depend on Genotypes" by Sandra Scarr,
Behavioral and Brain Sciences, March 1987.

100 "Her 'normal-range' concept": "Developmental Theories for the
Nineties," presidential address delivered at the Seattle meeting of the
Society for Research in Child Development, April 20, 1991.
"the environment can alter development": Ibid.
" 'Fortunately, evolution has not left development' ": Ibid.

101 " 'intellectual perversity' ": Interview with Sandra Scarr, July 7, 1993.
" 'It is no accident' ": *Race, Social Class and Individual Differences in
I.Q.*, edited by Sandra Scarr, Lawrence Erlbaum Associates, 1981.
"Politically she describes herself": Interview with Sandra Scarr, July 7,
1993.

102 " 'There is no more dangerous idea' ": *Race, Social Class and
Individual Differences in I.Q.*
" 'It is the suffering that should be addressed' ": Ibid.
"She found that the black children": "I.Q. Test Performance of Black
Children Adopted by White Families" by Sandra Scarr and Richard A.
Weinberg, *American Psychologist*, vol. 31, 1976.

103 "Scarr edited a book of essays": *Race, Social Class and Individual
Differences in I.Q.*

104 " 'The conclusion that we feel is justified' ": Ibid.

105 "she wrote a major statement on her overall views": "Developmental
Theories for the Nineties," presidential address delivered at the Seattle
meeting of the Society for Research in Child Development, April 20,
1991.
"first advanced by R. Q. Bell": "A Reinterpretation of the Effects in
Studies of Socialization" by R. Q. Bell, *Psychological Review*, vol. 75,
1968.

107 "(one by herself and Richard Weinberg)": "I.Q. Test Performance of
Black Children Adopted by White Families."

108 " 'the situations to which we want to generalize' ": *Race, Social Class
and Individual Differences in I.Q.*
" 'When I was about ten' ": Interview with Robert Plomin, November 17,
1993.

109 "A book he coauthored with his wife": *Why Children Are So Different*
by John Dunn and Robert Plomin, Basic Books, 1992.

111 "This genome-phenome-environmental interaction": *The Extended
Phenotype* by Richard Dawkins, W. H. Freeman, 1982.

111 "In his strongly antibehavior genetics book": *Biology As Ideology* by R. C. Lewontin, HarperPerennial, 1992.

112 " 'The environment includes whether you were lying' ": Interview with Dean Hamer, November 11, 1993.

113 "Plomin has written a basic textbook": *Behavioral Genetics: A Primer* by R. Plomin, J. C. Defries, and G. E. McClearn, W. H. Freeman and Company, 1980.
"Just fifteen years ago": "The Role of Inheritance in Behavior" by Robert Plomin, *Science*, April 13, 1990.

114 " 'will wonder what the nature-nurture fuss was all about' ": "Nature and Nurture" by Robert Plomin in *The Developmental Psychologists*, edited by M. R. Merrens and G. G. Brannigan, McGraw-Hill, 1995.
"he tried to preempt the usual debunking effort": "Overview: A Current Perspective on Twin Studies of Schizophrenia" by Kenneth Kendler, *American Journal of Psychiatry*, November 1983.

115 "He cited several studies": Ibid.
"He moved quickly to the main complaint": Ibid.
"identical twins who had been brought up": Ibid.

116 "As late as 1994": "Eugenics Revisited" by John Horgan, *Scientific American*, June 1993.

TEN: THE OTHER END — SEARCHING THE DNA

118 "Steve Jones has made do": *The Language of Genes* by Steve Jones, Anchor Books, 1994.

119 "Dani showed me a fingertip glob": Interview with Dani Reed, April 1994.

120 " 'It is not a major breakthrough' ": Interview with Arlen Price, April 26, 1977.

122 "Simon LeVay's image": "Homosexuality and Biology" by Chandler Burr, *Atlantic Monthly*, March 1993.

124 "Dawkins is particularly good at": *The Selfish Gene* by Richard Dawkins, Oxford University Press, 1976.
"As chromosomes divide": *Behavioral Genetics: A Primer* by R. Plomin, J. C. Defries, and G. E. McClearn, W. H. Freeman and Company, 1980.
"but a 1994 book": *The Beak of the Finch* by Jonathan Weiner, Alfred A. Knopf, 1994.

126 "a prominent molecular biologist": Interview with Dean Hamer, November 22, 1993.

129 "when a group headed by Janice Egeland": "Bipolar Affective Disorder

in the Older Order Amish" by Janice A. Egeland et al., *Nature*, November 16, 1989.

130 "a genetic basis for manic depression": "Genetic Linkage and Mental Disorders," *Biological Psychiatry*, July 1994.
"a gene on chromosome 11 for alcoholism": "Genetic Predisposition to Alcoholism" by K. Blum, E. P. Noble, *Alcohol*, Jan.–Feb. 1993.

132 "is suggested by a statement of Richard Dawkins": *The Selfish Gene*.
"Plomin sees hope": Interview with Robert Plomin, November 17, 1993.

ELEVEN: MOVING RIGHT ALONG THE DOUBLE HELIX

134 "a physiological link to a behavior had at last been found": "A Difference in Hypothalamic Structure Between Heterosexual and Homosexual Men" by Simon LeVay, *Science*, August 1991.

135 "the issue appeared to be resolved in 1991": "A Genetic Study of Male Sexual Orientation" by J. M. Bailey and R. C. Pillard, *Archives of General Psychiatry*, vol. 48, December 1991.

137 "In the resulting paper": "A Linkage Between DNA Markers on the X Chromosome and Male Sexual Orientation" by Dean Hamer et al., *Science*, July 16, 1993.

138 "reerupted with full fury in late 1994": *The Bell Curve* by Charles Murray and Richard Herrnstein, Free Press, 1994.

139 "The next important advance occurred in Holland": *New York Times*, October 22, 1993.

141 "the Berrettini study was announced": "Chromosome 18 DNA Markers and Manic-Depression Illness" by Wade Berrettini et al., *Proceedings of the National Academy of Sciences of Philadelphia*, July 1994.

142 "In his 1994 book": *Social Structure and Testosterone* by Theodore Kemper, Rutgers University Press, 1994.

143 "in experiments with rhesus monkeys": "Early Stress and Adult Emotional Reactivity in Rhesus Monkeys" by Stephen J. Suomi, *The Childhood Environment and Adult Disease*, John Wiley and Sons, Inc., 1991.

144 " 'experience can push genetic constitution around' ": "How We Become What We Are" by Winifred Gallagher, *Atlantic Monthly*, September 1994.
"A clearer, more manageable example": *New York Times*, November 2, 1993.

TWELVE: THE UPS AND DOWNS OF HUMAN NATURE

147 "One might as well ask": *Homicide* by Martin Daly and Margo Wilson, Aldine de Gruyter, New York, 1988.

148 "it was a result of a physical malfunction": *The Broken Brain* by Nancy Andreasen, Harper and Row, 1984.

149 "A. L. Wigan wrote a book entitled": *A New View of Insanity: The Duality of the Mind* by A. L. Wigan. Longman, Brown and Green, 1844.
"the physiological approach dominated the field": Ibid.

150 "Schopenhauer . . . defined his concept of 'will' ": *Schopenhauer and the Wild Years of Philosophy* by Rudiger Safranski, Harvard University Press, 1990.

151 "which launched his theory of evolution": *On the Origin of Species* by Charles Darwin, John Murray, 1859.

152 "Wallace deviated on this one major point": *Darwinism* by Alfred Russell Wallace, Macmillan, 1889.

154 "with the discovery of a Central American howler monkey": *African Genesis* by Robert Ardrey, Atheneum, 1961.
"Mendel's equally monumental findings were virtually ignored": *The Blind Watchmaker* by Richard Dawkins, W. W. Norton, 1987.

155 "But never before had such powerful notions": *The Broken Brain.*

156 "Nancy Andreasen, in her 1984 book": Ibid.

157 "In Europe, however, Freudianism": *In Search of Human Nature* by Carl Degler, Oxford University Press, 1990.
"The fifty-year triumph of Freudianism": Ibid.
"boost from Freud's disciples": *The Story of Psychology* by Morton Hunt, Anchor Books, 1993.

158 "Galton launched eugenics": *In Search of Human Nature.*
"launched a eugenics society in the United States": *In the Name of Eugenics* by D. Kevles, Penguin, 1986.

159 "Boas eventually became": *In Search of Human Nature.*

162 "in his exhaustively researched book": *Margaret Mead and Samoa* by Derek Freeman, Harvard University Press, 1983.

163 "Ruth Benedict published her major work": *Patterns of Culture* by Ruth Benedict, Houghton Mifflin, 1934.

165 "When eugenics top dog Charles Davenport could write": *In Search of Human Nature.*

166 "John B. Watson launched the behaviorist school": *The Story of Psychology.*

166 "Robert Plomin and his coauthors": *Behavioral Genetics: A Primer* by R. Plomin, J. C. Defries, and G. E. McClearn, W. H. Freeman and Company, 1980.
"Ada Yerkes, working with albino and normal rats": *In Search of Human Nature.*
167 "progressives pointed to the Social Darwinism": Ibid.
168 "Lysenko adopted the more effective strategy": *The Rise and Fall of T. D. Lysenko* by Z. A. Medvedev, Columbia University Press, 1969.
169 "This sorry tale of his rise and fall": *Heredity East and West: Lysenko and World Science* by J. Huxley, Shuman, 1949.

THIRTEEN: SHORT AND HAPPY LIFE OF THE TABULA RASA

171 "a conference entitled 'Genetics and Social Behavior': *In Search of Human Nature* by Carl Degler, Oxford University Press, 1990.
172 "The first counterdevelopment": *King Solomon's Ring* by Konrad Lorenz, Crowell, 1952; and *The Study of Instinct* by N. Timbergen, Oxford University Press, 1951.
"reeducated animals invariably returned to the behavior": *The Story of Psychology* by Morton Hunt, Anchor Books, 1993.
173 "Harlow's experiments involved infant monkeys": Ibid.
174 "The book that resulted": *African Genesis* by Robert Ardrey, Atheneum, 1970.
176 "In a later book": *The Social Contract* by Robert Ardrey, Atheneum, 1970.
177 "In *The Third Chimpanzee*": *The Third Chimpanzee* by Jared Diamond, HarperPerennial, 1993.
178 "a 1920 work by an English bird-watcher": *Territory in Bird-Life* by Eliot Howard, William Collins, 1920.
"an American zoologist": *Behavior and Social Relations of the Howling Monkey* by C. R. Carpenter, John Hopkins Monographs in Comparative Psychology, 1934.
"in his 1952 book": *King Solomon's Ring.*
179 "a number of ethological books": *On Aggression* by Konrad Lorenz, Methuen, 1966; *The Naked Ape* by Desmond Morris, Constable, 1967; *Men in Groups* by Lionel Tiger, Random House, 1969; *The Imperial Animal* by Robin Fox and Lionel Tiger, Holt, Rinehart and Winston, 1971.
180 "Ardrey's next book": *Territorial Imperative* by Robert Ardey, Atheneum, 1966.

180 "by the near simultaneous publication of": *On Aggression.*
"When I interviewed Lionel Tiger": Interview with Lionel Tiger,
June 4, 1997.

FOURTEEN: SURVIVING THE JENSEN FUROR

183 "abruptly derailed in 1969": "How Much Can We Boost I.Q. and
Scholastic Achievement?" by Arthur Jensen, *Harvard Educational
Review*, Winter issue, 1969.
"the same debate erupted again": *The Bell Curve* by Charles Murray
and Richard Herrnstein, Free Press, 1994.

184 "I.Q. was seen as another device": *The Mismeasure of Man* by Stephen
Jay Gould, W. W. Norton, 1981.

185 "and led the pack in denouncing": The best summary of these critics is
in the book edited by Sandra Scarr, *Race, Social Class and Individual
Differences in I.Q*, edited by Sandra Scarr, Lawrence Erlbaum
Associates, 1981.

186 Interview with Robert Plomin, November 17, 1993.

187 "Not only are Afro-Americans": *Eco Homo* by Noel Boaz, Basic Books,
1997.

188 "the publication in 1975": *Sociobiology: A New Synthesis* by
Edward O. Wilson, Harvard University Press, 1975.

189 "and drew parallels with racism and Nazism": *The Naturalist* by
Edward O. Wilson, Island Press, 1994.

191 "Wilson himself, who said a 1983": *Promethean Fire: Reflections on the
Origins of Mind* by E. O. Wilson and C. J. Lumsden, Harvard
University Press, 1983.
"opened people's eyes to the chemical basis": *The Broken Brain* by
Nancy Andreasen, Harper and Row, 1984.

192 "Freudians were further debilitated": *The Rise and Crisis of
Psychoanalysis in the United States* by Nathan G. Hale Jr., Oxford
University Press, 1997.
"Books denouncing psychoanalysis appeared": *Final Analysis* by J. M.
Masson, Addison-Wesley, 1990.

193 "by Frederick Crews": *New York Review of Books*, November 18, 1993.
"44 percent of those polled": *Consumer Reports*, November 1995.

194 "*The New York Times* ran a cover story": *New York Times*, October 16,
1994.
"*The New Republic* ran an article": "Race, Genes and I.Q.: An
Apologia," *The New Republic*, October 31, 1994.

194 " 'an Orwellian ritual-denunciation session' ": *Wall Street Journal*, October 20, 1994.

195 "an ethnic group called the Burakumin": "Race Genes and I.Q.," by Ned Block, *Boston Review*, December/January, 1995–96.
"Research along these lines": "A Threat in the Air" by Claude M. Steele, American Psychologist, vol. 52, 1997.

FIFTEEN: OH SO POLITICAL SCIENCE

198 "A few years later, Richard Herrnstein": "I.Q." by Richard Herrnstein, *Atlantic Monthly*, September 1971.
"a book that set out to deconstruct": *Not in Our Genes* by R. C. Lewontin, L. J. Kamin, and S. Rose, Pantheon Books, 1984.

203 "he mentioned this as a problem": Interview with Jonathan Beckwith, September 21, 1993.

204 "In a paper criticizing the Minnesota Twin Study": "The Genetic Analysis of Human Behavior: A New Era?" by J. Beckwith, P. Billings, and J. S. Alper, *Social Science and Medicine*, vol. 35, 1992.

206 " 'Behavioral genetics gets people excited because it is important' ": Interview with Matthew McGue, November 13, 1993.
"He wistfully cites an illustration": Interview with Robert Plomin, November 17, 1993.
" 'Yes, we are complaining' ": "Having the Last Word" by Sandra Scarr, *Race, Social Class and Individual Differences in I.Q.*, edited by Sandra Scarr, Lawrence Erlbaum Associates, 1981.

207 "Hamer next got a letter": Interview with Dean Hamer, November 22, 1993.

209 " 'I wasn't sure if I should kiss Lewontin's ring' ": *The Science of Desire* by Dean Hamer and Peter Copeland, Simon and Schuster, 1994.

210 " 'Standards of evidence are raised' ": Interview with Sandra Scarr, July 7, 1993.

212 "two books appeared": *The Burt Affair* by Robert B. Joynson, Routledge, 1989; and *Science, Ideology and the Media* by Ronald Fletcher, Transaction Publishers, 1991.

213 "which is the central theme of": *The Mismeasure of Man* by Stephen Jay Gould, W. W. Norton, 1981.

214 " 'I had no specific scientific purpose in mind' ": Interview with Thomas Bouchard, November 12, 1993.

215 "alleges cooperation with the Nazis": *Konrad Lorenz: The Man and His Ideas* by Richard I. Evans, Harcourt Brace Jovanovich, 1975.

Notes

SIXTEEN: SCIENTISTS IN DENIAL

216 "In the epilogue to a book": *Behavior Genetic Analysis*, edited by
Gerald Hirsch, McGraw-Hill, 1967.

217 "a forceful exponent of Hirsch's views": Interview with Tim Tully,
August 2, 1993.
"Tully cites a 1907 experiment": Ibid.

218 "could not resist a swipe": Ibid.

219 "Hubbard wrote with exasperation in a letter": *The New Yorker*,
April 25, 1994.

220 " 'We can't control the weather' ": Interview with Thomas Bouchard,
November 14, 1993.
"Anthropologist Robin Fox": *The Challenge of Anthropology* by Robin
Fox, Transaction Publishers, 1994.
" 'They're crooks!' ": Interview with James Watson, August 2, 1993.
"In *Myths of Gender*": *Myths of Gender* by Ann Fausto-Sterling,
HarperCollins, 1985.
"Ruth Hubbard, in her 1993 book": *Exploding the Gene Myth* by
Ruth Hubbard and Elija Wald, Beacon Press, 1993.

221 " 'Of course there is a genetic component' ": Interview with Leon
Kamin, September 22, 1993.

222 "A standout example was in a review": *New York Review of Books*,
April 7, 1994.

223 " 'I wish' ": Interview with Sandra Scarr, May 6, 1993.
"A culmination of sorts": *Time*, August 15, 1994.
" 'It's time to move on' ": Interview with David Lykken, November 15,
1993.
"an article that purported to be": "Eugenics Revisited" by John
Horgan, *Scientific American*, June 1993.

225 "One of the few letters published": *Scientific American*, November 23,
1993.

SEVENTEEN: VIOLENT CRIME IN THE MARYLAND WOODS

229 "began in 1870 with Cesare Lombroso": *Crime and Human Nature*
by James Q. Wilson and Richard J. Herrnstein, Touchstone Books,
1985.
"and published it in an influential book": *The Delinquent Man* by
C. Lombroso, F. Alcan, 1887.

230 "the histories of 709 members of an Irish family": *The Jukes: A Study in*

Crime, Pauperism, Disease and Heredity by R. L. Dugdale, G. P. Putnam and Sons, 1910.

230 "Englishman Charles Goring": *Crime and Human Nature* by James Q. Wilson and Richard J. Herrnstein, Touchstone Books, 1985.

"whose generations of retards inspired the eugenics movement": *The Kallikak Family: A Study in the Heredity of Feeblemindedness* by H. H. Goddard, Macmillan, 1912.

"Virginia's notorious Lynchburg Colony": *The Lynchburg Story* (documentary film), directed by Bruce Eadie, Worldwide Pictures, 1994.

231 "Germany joined other European nations": *In Search of Human Nature* by Carl Degler, Oxford University Press, 1990.

"most notably with William Sheldon's theories": *The Story of Psychology* by Morton Hunt, Anchor Books, 1993.

233 "had announced a 'violence initiative' ": "The Biology of Violence" by Robert Wright, *The New Yorker*, March 13, 1995.

"who in a speech": Ibid.

234 "A Washington newspaper": Ibid.

236 "animal breeders who for centuries": *On Aggression* by Konrad Lorenz, Harcourt Brace, 1966.

"neurologists and clinical psychologists had long known": "Neurology of Aggression and Episodic Dyscontrol" by F. A. Elliott, *Seminars in Neurology*, vol. 10, no. 3, September 1990.

"a small section of the hypothalamus": Ibid.

"The most impressive study": *The Causes of Crime: New Biological Approaches* by S. A. Mednick, T. E. Moffitt, and S. A. Stack, Cambridge University Press, 1987.

237 "In a 1990 paper Elliott wrote": "Neurology of Aggression and Episodic Dyscontrol" by F. A. Elliott, *Seminars in Neurology*, vol. 10, no. 3, September 1990.

238 "about one-eighth of the population": Interview with David Lykken, November 15, 1993.

239 "the landmark book": *Homicide* by Margo Wilson and Martin Daly, Aldine de Gruyter, 1988.

242 "when a Texas child molester": Associated Press, April 8, 1996.

247 "a book called": *All God's Children* by Fox Butterfield, Alfred A. Knopf, 1995.

250 "the sexual abuse of a retarded girl": *Our Guys* by Bernard Lefkowitz, University of California Press, 1997.

251 "by shutting off a gene": *Nature*, November 23, 1995; and *New York Times*, November 23, 1995.

"that human hemoglobin": *New York Times*, March 21, 1996.

252 "a connection existed between": "Bone Lead Levels and Delinquent Behavior" by H. L. Needleman et al. *Journal of the American Medical Association*, February 7, 1996.

EIGHTEEN: CONCLUSIONS

256 "foresee the possibility of raising I.Q.": *Race, Social Class and Individual Differences in I.Q.*, edited by Sandra Scarr, Lawrence Erlbaum Associates, 1981.

257 "and learned of the hormonal mishaps": "Homosexuality and Biology" by Chandler Burr, *Atlantic Monthly*, March 1993.
"Ashkenazi Jews": *New York Times*, December 7, 1993.

259 " 'The shortcomings of our description' ": *Beyond the Pleasure Principle* by Sigmund Freud, Boni and Liveright, 1920.

261 "As Robert Wright said": *The Moral Animal: Evolutionary Psychology and Everyday Life* by Robert Wright, Pantheon, 1994.
"Some evolutionary thinkers": *Adaptation and Natural Selection* by George Williams, Princeton University Press, 1966.

263 "Eysenck did a study": *The Psychology of Politics* by H. J. Eysenck, Routledge & Keegan Paul, 1954.
"Other studies have indicated ": "Transmission of Social Attitudes" by N. G. Martin et al., *Protocol of National Academy of Science*, June 1986; and "The Importance of Heritability in Psychological Research: The Case of Attitudes" by Abraham Tesser, *Psychological Review*, vol. 100, 1993.

267 "Startling research on brain function": *Nature's Mind* by Michael Gazzaniga, Basic Books, 1992.

269 "the speed with which humans": *Demonic Males* by Richard Wrangham and Dale Peterson, Houghton Mifflin, 1996.

270 " 'I must fight til' ": by Marianne Moore, *The Marianne Moore Reader*, Viking, 1961.

271 "One of the best analyses": *Love Thy Neighbor* by Peter Maass, Alfred A. Knopf, 1996.

272 "wars are no longer waged": *Civil Wars* by Hans Magnus Enzenberger, The New Press, 1994.
"who periodically attack their neighbors": *The Third Chimpanzee* by Jared Diamond, HarperPerennial, 1993.

Index

abortion, 257–59, 264–65, 266
addictions, 14, 16, 255–56, 257, 267. *See also* alcoholism
adoption studies, 7–8, 37, 85, 198; criticisms of, 205–9, 217, 221; and on intelligence, 88, 102–3, 107, 110, 221; methodology of, 205–9; and personality, 49–50, 87–88; and race, 102–3
African Genesis (Ardrey), 174–80
age, 39, 56, 68–69, 88
aggression, 53, 55, 139–40, 174–80, 236, 269–72. *See also* violence; Wye Conference
alcoholism, 9, 18, 94, 95, 97, 130–31, 187, 208, 222, 224–25, 242
All God's Children (Butterfield), 247–51
American Psychologist, 188, 275–78
Amish families depression study, 129–30, 131, 133, 141, 142, 224
Andreasen, Nancy, 156
Angier, Natalie, 239
animals: aggression in, 236; and behaviorism, 166; evolution of behavior in, 35; family life of, 144–46; and gene-environment interaction, 143–44; genetics as basis of behavior in, 11; human studies compared with, 130; humans as, 152–55, 224; 1950s and 1960s

studies of, 172–74; and prairie voles study, 144–46; and sociobiology, 188–89, 190; specific behavioral gene in, 10
Ardrey, Robert, 174–80, 181, 189, 269
astrology, 73
attitudes and genes, 56, 262–67
Australian studies, 263–64

backward-bathing twins, 79–80, 84
Bailey, Michael, 51, 135–36
Baker, Russell, 18
Balaban, Evan, 207–8, 209, 238–39
Baron, Miron, 130
Beckwith, Jonathan, 203–5, 234, 238
beer-drinking firemen, 77
behavior: definition of, 5–6
behavioral genetics: acceptance of, 13, 17–18, 45, 85, 197, 223, 238; and "bad" knowledge, 215; benefits/misuses of, 7, 17, 255–73; emergence of, 182; as fad, 16, 193; as failure, 129–32, 133, 140–41; multigene aproach to, 131–33; observational studies compared with, 9–11; and specific behavioral genes, 8–9, 122–23, 128–33, 139, 239, 242; summary of state of, 9. *See also specific researcher, study, or topic*

295

Index

Dudley, Richard, 58, 59
Dugdale, Robert L., 230
Dunn, Judy, 108–9
Dutch research, 134, 135, 139–40
DZ. *See* fraternal twins

Egeland, Janice, 129–30, 131
Elliott, Frank, 237
emergenesis theory, 83–84
Ende, Robert, 87
environment: adverse conditions in,
    60–61; and complexity of behavior, 255;
    complexity of, 110–13; control of, 219–20;
    definition of, 4, 112; discrediting of
    domination of, 42–45, 100–108; as
    dominant in explaining behavior, 11–17;
    effects on internal biochemistry of, 143;
    and evolution of behavior, 6–7, 35;
    extremes in, 100; measurement of, 217;
    misconceptions about, 16; normal range
    of, 60–61, 143, 144, 218, 219; and politics,
    12–17, 168, 169; shared/nonshared, 43, 45,
    96, 105–6, 109–10. *See also* behaviorism;
    culture; "customized" environment;
    Freudianism; gene-environment
    interaction; sociobiology; *specific
    researcher*
Enzenberger, Hans Magnus, 272
epistasis process, 83
ethnicity, 195–96
ethology, 174–82, 188
eugenics, 13, 158–59, 160, 165, 223, 226,
    230–31, 243, 257–59
evolution: and academic disciplines, 190;
    of behavior, 6–7, 18–19, 35, 57–58; and
    brain chemistry, 93–94; Darwin's work
    on, 151–55; and environmental
    determinism, 6–7; and eugenics, 158;
    and mutations, 124–25; Plomin's views
    on, 108–9; poll about belief in, 73; and
    religion, 152
evolutionary psychology, 190, 193, 246–47,
    261, 273
eye experiments, 217–18
Eysenck, H. J., 262–63

family life, animal, 144–46
Farber, Susan, 23, 49

Fausto-Sterling, Ann, 220
finger ridges, 56
Fishbein, Diana, 240, 242
Fletcher, Ronald, 212, 213
Fossey, Dian, 174
Fox, Robin, 179, 190, 220
Fox, Sydney W., 48–50
fraternal twins, 39, 75–76, 115, 205; absence
    of similarities in, 78, 82–83, 84;
    behavioral inhibition of, 87; and
    coincidence explanation, 78; DNA of,
    21, 22; and homosexuality, 50, 51, 135–36;
    identical twins compared with, 7–8, 21,
    22, 73, 85, 114–17, 201, 236; intelligence
    of, 48–49, 87, 201; mental disorders of,
    96, 114–17; and multimind theory, 82–83;
    temperament of, 87, 88. *See also specific
    study*
fraud, 214–15. *See also* Burt, Cyril
Freeman, Derek, 162, 163
Freeman, F. N., 22, 80–81
Freud, Sigmund, 50, 77, 94, 150, 210, 259
Freudianism, 130, 155–58, 166, 170, 171,
    192–93
fruit fly experiments, 217

Galton, Francis, 21, 158
Gazzaniga, Michael, 81–82, 267–69
gender issues, 50–51, 96, 187, 199, 237–38
gene-environment interaction: complexity
    of, 200, 219; and environment's impact
    on genes, 60–61, 196, 217–20; and
    genetics' influence on environment,
    43–44, 196; and *Not in Our Genes*
    criticisms, 200; and parents' genetics,
    106–7, 110–11; and prairie vole study,
    142–44
genes: complexity of, 133, 239; definition
    of, 126–27; as dictators of behavior, 14–15,
    220–21, 254; and DNA of twins, 21–22;
    dynamism of, 239; as enemies, 261–62;
    markers on, 128–29, 232; measurement
    of influence of, 221; mechanics/
    functions of, 127–28; overview of, 123–28;
    and physical traits, 5, 7, 14, 56, 64, 67,
    69, 86, 131; and polymerase chain
    reaction of DNA, 118–19; and selecting a
    trait for search, 119–20; and sibling

Index

Mendelian genetics, 131, 151, 154–55, 168, 169

mental disorders: of Amish families, 129–30, 131, 133, 141, 142; and benefits/misuses of behavioral genetics, 187, 256, 259, 267; changing beliefs about, 133, 148–50; of fraternal twins, 114–17; and Freudianism, 155–56; of identical twins, 79–80, 114–17; and mood-altering drugs, 191–92; of reared-apart twins, 79–80; of reared-together twins, 94–98; specific gene for, 267; testing for, 150. *See also* depression; phobias; schizophrenia; *specific researcher*

methodology, 85, 205–9, 212. *See also specific researcher or study*

mice studies, 251–52

migraine headaches, 76

Mill, John Stuart, 150

Minnesota Twin Study: and background information, 39; and benefits/misuses of behavioral genetics, 262, 263; case reports from, 63–67; credibility of, 41, 42; criticisms of/opposition to, 41, 42–45, 58–61, 201–2, 203–5, 214, 221, 224–27; deliberate distortion/fraud in, 72, 214–15; findings from, 8, 32–33, 35, 40, 48–51, 52–61, 63–76, 77–84, 97–98, 135, 198, 205, 236, 262, 263; funding for, 36–38, 215; goal of, 34; Goleman's comments about, 54–55; and heritability mean figures of concordance, 79; impact/significance of, 42, 62–63, 76, 224; informed-consent agreement of, 56, 215; and Jensen furor, 187–88; Jim Twins as impetus for, 25–26, 28–32; and media, 73–74, 85, 98, 198; methodology/planning for, 34–36, 56, 58–59, 85, 201–2, 205, 214; and politics, 198; publications from, 42, 50–51, 54, 55–58, 62; recruitment of twins for, 25, 26, 29, 32, 37–39, 70–71; and replication studies, 38, 97–98, 116, 213–14; sample size/selection for, 8, 41, 52, 79, 141; scope of, 141; staff for, 39; testing for, 39–41. *See also* Bouchard, Thomas

Mischel, W., 33

molecular genetics, 113–14, 121–23, 132. *See also specific researcher*

monkey studies, 218, 233

monozygotic twins (MZ). *See* identical twins

morality, 262–67

Morris, Desmond, 179

Moynihan, Daniel Patrick, 198

multimind theory, 81–83

Murray, Charles, 138–39, 183, 193–94, 263–64

mutations, 124–25

MZ. *See* identical twins

nail biting, 31

National Institutes of Health, 87, 130, 135, 144–46, 232, 234

National Science Foundation, 46

natural selection, 13, 152, 154, 188, 261

Neubauer, Peter, 11–12

Newman, H. H., 22, 80–81

nitric oxide, 251–52

Nobel, Ernest, 130–31

nonshared environment, 43, 45, 105–6, 110

norepinephrine, 143, 144

"normal-range" concept, 60–61, 100, 106, 107, 143, 144, 147, 218, 219

Not in Our Genes (Lewontin et al.), 198–202, 220

obesity, 10, 18, 119–21

observational studies, 9–11, 113, 119, 121. *See also specific researcher or study*

occupations, 33, 39

oxytocin, 144–46

Page, George, 17

parenting, 12, 57, 103–7, 115–16, 259

Parker, Barbara, 36

Paul, Diane, 238–39

personality: and adoption studies, 49–50, 88; and age, 88; and astrology, 73; early studies about genetics and, 97, 101; and environment, 42–43, 49–50, 53, 54, 143; of fraternal twins, 73; Freud's views about, 155–58; and gene-environment

Index

University of Minnesota Twin Study. *See* Bouchard, Thomas; Minnesota Twin Study
University of Pennsylvania, 118–21
University of Pittsburgh, 252
U.S. Naval Academy midshipmen and testosterone, 143

van der Post, Laurens, 156
vasopressin, 144–46
violence: and "bad blood," 229; and benefits/misuses of behavioral genetics, 14, 16, 187, 222, 256, 257, 269–72; chemical links to, 251–53; and culture, 247–51; definition of, 234; and drug therapy, 233, 236, 240, 241, 242, 243, 245, 247, 253, 256; and lead, 252; and nitric oxide, 251–52; and politics, 233–35; and poverty, 243, 244, 249; and race, 233–34, 238, 243, 244, 246, 247–51; and reared-apart twins, 236, 239; and specieswide differences, 246–47; specific gene for, 139–40, 222, 239, 242. *See also* aggression; criminals; Wye Conference

Wald, Elijah, 220
Wallace, Alfred, 152, 190
Wasserman, David, 232–35, 238, 239, 244
Watson, James, 172, 202, 220, 222

Watson, John B., 50, 89, 166–67, 169–70, 172–73, 178, 197, 210
Weinberg, Richard, 102–3, 104, 107
Weiner, Jonathan, 125
Wexler, Nancy, 129
Weyher, Harry, 47
Wigan, A. L., 149
Williams, George, 261
Wilson, Edward O., 35, 45, 101, 188–90, 191
Wilson, Margot, 147–48
Wilson, R. S., 86–87
Wilson, Victoria, 49
Wright, Robert, 261
Wye Conference (1995): aim of, 232, 246; background about, 228, 232–35; conclusions of, 246–47; funding for, 232, 235–36; and media, 233, 245–46; and politics, 233–34, 235–36; protesters at, 243–46, 253; speakers/participants at, 238–43

Yerkes, Ada, 166
Yufe, Jack, 67–72, 74, 75, 77, 84, 203–4
YY chromosome, 231, 232

Zahn-Waxler, Carolyn, 87
Zimring, Frank, 241
zygosity: determination of, 39, 115

A NOTE ON THE TYPE

The text of this book was set in Electra, a typeface designed by W. A. Dwiggins (1880–1956). This face cannot be classified as either modern or old style. It is not based on any historical model, nor does it echo any particular period or style. It avoids the extreme contrasts between thick and thin elements that mark most modern faces, and it attempts to give a feeling of fluidity, power, and speed.

Composed by Creative Graphics, Allentown, Pennsylvania
Printed and bound by Quebecor Printing, Fairfield, Pennsylvania
Designed by Robert C. Olsson

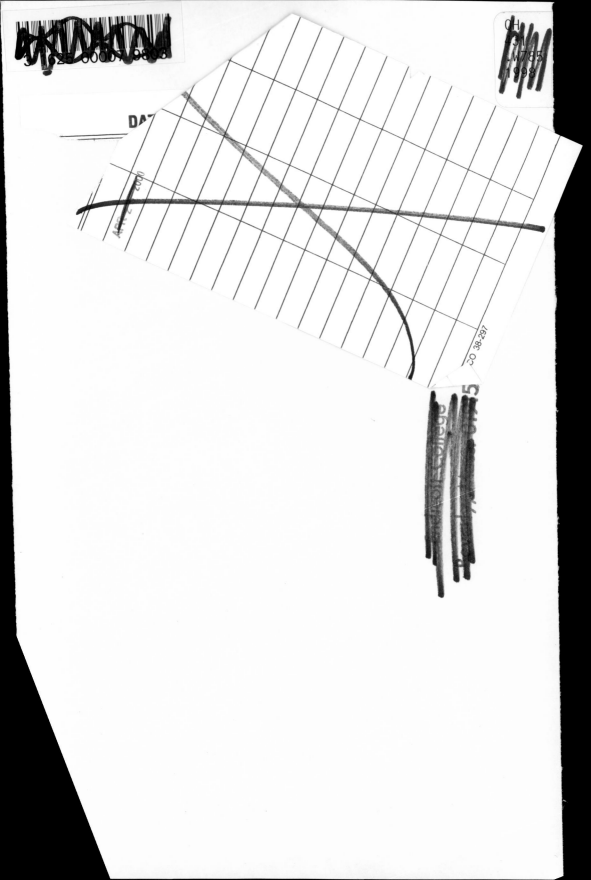